Fluid Mechanics and Its Applications

Founding Editor

René Moreau

Volume 138

The purpose of this series is to focus on subjects in which fluid mechanics plays a fundamental role. As well as the more traditional applications of aeronautics, hydraulics, heat and mass transfer etc., books will be published dealing with topics, which are currently in a state of rapid development, such as turbulence, suspensions and multiphase fluids, super and hypersonic flows and numerical modelling techniques. It is a widely held view that it is the interdisciplinary subjects that will receive intense scientific attention, bringing them to the forefront of technological advancement. Fluids have the ability to transport matter and its properties as well as transmit force, therefore fluid mechanics is a subject that is particularly open to cross fertilisation with other sciences and disciplines of engineering. The subject of fluid mechanics will be highly relevant in such domains as chemical, metallurgical, biological and ecological engineering. This series is particularly open to such new multidisciplinary domains. The median level of presentation is the first year graduate student. Some texts are monographs defining the current state of a field; others are accessible to final year undergraduates; but essentially the emphasis is on readability and clarity.

Springer and Professor Thess welcome book ideas from authors. Potential authors who wish to submit a book proposal should contact Dr. Mayra Castro, Senior Editor, Springer Heidelberg, e-mail: mayra.castro@springer.com

Indexed by SCOPUS, EBSCO Discovery Service, OCLC, ProQuest Summon, Google Scholar and SpringerLink

Helen Wilson · Sarthok Sircar · Priyanka Shukla

Viscoelastic Subdiffusive Flows

Theory and Computation

Springer

Helen Wilson
Department of Mathematics
University College London
London, UK

Sarthok Sircar
School of Mathematical Sciences
Indraprastha Institute of Information Technology Delhi
New Delhi, Delhi, India

Priyanka Shukla
Department of Mathematics
Indian Institute of Technology Madras
Chennai, Tamil Nadu, India

ISSN 0926-5112 ISSN 2215-0056 (electronic)
Fluid Mechanics and Its Applications
ISBN 978-981-96-0134-9 ISBN 978-981-96-0135-6 (eBook)
https://doi.org/10.1007/978-981-96-0135-6

Mathematics Subject Classification: 76A10, 76E17, 53Z05, 53C21, 47D03, 26A33

This Springer imprint is published by the registered company Springer Nature Singapore Pte Ltd.
The registered company address is: 152 Beach Road, #21-01/04 Gateway East, Singapore 189721, Singapore

Fractional calculus finds use in many fields of science and engineering, including fluid flow, rheology, diffusive transport akin to diffusion, electrical networks, electromagnetic theory, and probability. It seems that hardly a field of science or engineering has remained untouched by this topic. Yet even though the subject is old, it is rarely included in today's curricula. Possibly, this is because many mathematicians are unfamiliar with its uses...

—K. S. Miller and B. Ross.

Preface

This book explores the dynamical and rheological properties of high molecular weight entangled polymer melts, the so-called viscoelastic sub-diffusive fluids, through a combination of linear stability analysis, differential geometric construction, and computational simulations. Unlike dilute polymeric liquids, the molecular transport of these materials occurs within the sublinear regime (i.e., the timescale of the diffusive transport is $t^{\alpha}, 0 < \alpha \leq 1$). The potential of fractional calculus is highlighted in modelling these types of fluids. A novel fractional model is developed to investigate regions of spatiotemporal instability of channel flows. The entanglement of the polymer chains at the microscale leads to localized, non-homogeneous regions with higher viscosity, the so-called 'spatiotemporal macrostructures'. Direct numerical simulations are utilized to capture these macrostructures within the flow, using a newly constructed structure tensor through the scalar invariants of a Riemannian metric, and thus contributing to a deeper understanding of this class of complex fluids. The description of the five chapters in the book is as follows.

Chapter 1 introduces complex fluids, specifically polymeric fluids, and delves into the many strategies used to measure and model them. The motivation for considering sub-diffusive flows is introduced, along with an introduction to diffusive motions and anomalous diffusion.

Chapter 2 lays the fundamental groundwork for the various kinds of temporal and spatiotemporal instability occurring in this class of flows: convective instabilities, absolute instabilities, and evanescent modes. We embark upon a comprehensive stability investigation of the viscoelastic sub-diffusive plane Poiseuille flow. A brief backdrop of fractional models applied to viscoelastic flows is presented. The Fractional Upper Convected Maxwell (FUCM) model is developed and explored under low to moderate Reynolds and Weissenberg numbers to uncover regions of topological transition of the flow interface. The interrelation between the exponent of the power-law scaling in microscale models (t^{α}) and fractional order (α) in stress constitutive equations is investigated. Using Brigg's method of analytic continuation, insights into instability modes, along with the changes in the fractional time derivative order, are discussed. The stability studies for two well-renowned models are presented: monomer diffusion in Rouse chain melts, $\alpha = 1/2$, and in Zimm chain solutions, $\alpha = 2/3$.

Chapter 3 presents a theory to quantify and visualize the development of spatiotemporal macrostructures for viscoelastic sub-diffusive flows by decomposing the polymer conformation tensor into the so-called structure tensor. Our method gets over the traditional arithmetic decomposition's fundamental flaw: fluctuating conformation tensor fields might not be positive definite and, as a result, lose their physical significance. By defining and building a geodesic via the inner product on its tangent space, the space of positive definite matrices is converted into a Riemannian manifold using well-proven results in matrix analysis. Three scalar invariants of the structure tensor are defined through this geodesic.

Chapter 4 describes the direct numerical simulations of viscoelastic, sub-diffusive, plane Poiseuille flow in low to moderate Reynolds number and low Weissenberg number regime. A family of time-asymptotically stable, implicit-explicit, adaptive time integration methods (denoted with the θ-method) for the solution of the fractional advection-diffusion-reaction (FADR) equations is developed. A generalized version of the computationally explicit L_1 method is introduced via this class of temporal integration technique. We highlight how these simulations successfully capture the flow structures by offering (i) a better resolution of the instantaneous regions of elastic shocks (which are the alternating regions of expanded and compressed polymer volume, in comparison with the volume of the mean conformation tensor) and (ii) a higher resolution to identify areas where the mean conformation tensor tends to diverge significantly from the instantaneous conformation tensor, thus supporting an experimentally observed instability transition of sub-diffusive flows.

Chapter 5 has been added to present a background for some unsolved problems involving these fascinating materials because of the rapidly expanding interest in viscoelastic polymeric melts. The rheological properties of polymeric melts involving real-life applications are summarized. This chapter concludes by highlighting the fundamental challenges in quantifying these flows as well as the potential future problems demanding serious attention in this new class of complex fluids.

Finally, as educators in the area of modelling sub-diffusive flows, the authors want to address the group of graduate students. The standard knowledge acquired during usual university education in mathematics, physics, and engineering does not enable a quick start of research activity in sub-diffusive flows. Indeed, in this typically interdisciplinary field, everyone can lay emphasis on matters that are familiar to them because of their preceding education. Significant research activities, however, can only be generated based on more universal knowledge. In the absence of this knowledge, beginners have to rely upon the guidance of their supervisors for an unduly long period. Otherwise, they take the risk of losing too much of their costly time.

Hence, this book is an attempt to abridge this troublesome path towards the research frontier of sub-diffusive flows/polymeric melts to enable the younger generation to execute research within a reasonable time frame.

London, UK
New Delhi, India
Chennai, India
December 2023

Helen Wilson
Sarthok Sircar
Priyanka Shukla

Acknowledgements

The authors owe an immeasurable debt of thanks to many of their colleagues and mentors through whom they have been educated in this field. Professor Wilson expresses her gratitude to her family for putting up with the efforts associated with co-authoring a book at the end of five long years as Head of Department (and all the lockdowns). Special thanks go to the polymeric fluids group from the early 2000s at the University of Leeds—Dr. Oliver Harlen, Prof. Daniel Read, Prof. Suzanne Fielding, and, of course, the late Prof. Tom McLeish, who nurtured her in her early career and introduced her to the area of polymer melts.

Dr. Sircar wants to express his gratitude to Prof. Ganesh Subramanian, Prof. Abhijit Deshpande, Prof. Susy Varugesse and Prof. Manoranjan Mishra, due to their suggestions in various stages of research leading to the eventual writeup of the chapters, as well as giving their valuable time to discuss the pitfalls associated with the modelling content. Special mention goes to Prof. Jeff Giacomin for introducing the authors to the potential of fractional calculus in modelling polymer melt rheology. Dr. Sircar is indebted to his coworker, Dr. Kaushik Kalyanaraman, for finding the proofs of the theorems related to the geodesic construction on the Riemannian manifold, as well as to his graduate students, Tanisha Chauhan and Madhu Bhatt for their aid in preparing the figures. In addition, the authors are indebted to the Department of Science and Technology, India and their respective host institutes and their management, since without their wholehearted support of basic research in general and this work in particular, this volume would certainly not exist.

Dr. Shukla expresses her deepest gratitude to Prof. Meheboob Alam, Prof. Anne De Wit, Prof. R. Usha, and Prof. Vicente Garzó for many fruitful discussions on complex fluids and for providing continuous academic support. Dr. Shukla also thanks her Ph.D. students, Dr. Lima Biswas, Dr. Sreethin Sreedharan K., Dr. Supriya Karmakar and Yudhvir Singh Yadav, for their help and brain-storming research sessions. Dr. Shukla sincerely acknowledges financial support from the DST and the ICSR, IIT Madras. Dr. Shukla dedicates this book to her family, in particular, to her husband, Vinay, and son, Gaurang, for unwavering support and understanding during the demanding phases of the research.

Competing Interests The authors have no conflicts of interest to declare that are relevant to the content of this book.

Dr. Helen Wilson acknowledges the support of EPSRC over many years via grants GR/T11807/01 and EP/N024915/1.

Dr. Sarthok Sircar acknowledges the financial support of DST CRG/2022/000073.

Dr. Priyanka Shukla acknowledges the financial support of grants MTR/2023/000892, CRG/2023/002914 of DST, and from the ICSR, IIT Madras.

Contents

Acronyms and Symbols

ω_i^{Temp}	Temporal Growth Rate
ω_i^{Cusp}	Absolute Growth Rate
V_g	Group Velocity
Re	Reynolds Number
We	Weissenberg Number
$\mathbf{v}$	Fractional Velocity
$\dot{\gamma}$	Shear Rate
λ^α	Relaxation Time
G	Elastic modulus
$G(t)$	Relaxation Spectrum
G'	Storage modulus
G''	Loss modulus
ρ	Density
p	Isotropic Pressure
τ	Elastic Stress
η_s	Solvent Viscosity
η_p	Polymeric Contribution to Shear Viscosity
η_0	Total Viscosity $(= \eta\text{s} + \eta\text{p})$
ν	Viscous Contribution to the Total Viscosity
$\boldsymbol{A}$	Polymer Stress Tensor
$\boldsymbol{D}$	Rate of Strain Tensor
$\boldsymbol{L}$	Velocity Gradient Tensor
Ω	Vorticity
ψ	Streamfunction
$\mathcal{C}$	Instantaneous Conformation Tensor
$\overline{\mathcal{C}}$	Mean Conformation Tensor
$\mathcal{C}'$	Fluctuating Conformation Tensor
$\mathcal{G}$	Instantaneous Structure Tensor
$\mathcal{G}'$	Fluctuating Structure Tensor
$\boldsymbol{F}$	Instantaneous Deformation Gradient Tensor

$\overline{\boldsymbol{F}}$	Mean Deformation Gradient Tensor
$\mathcal{L}$	Fluctuating Deformation Gradient Tensor
$\boldsymbol{GL}_n$	Set of all n × n Invertible Matrices
$\boldsymbol{SO}_n$	Set of all n × n Special Orthogonal Matrices
$\boldsymbol{PS}_n$	Set of all n × n Positive Definite Matrices
C^m	Set of m−times continuously differentiable real valued functions
$\mathcal{C}$	Conformation Tensor
N_c	CFL Number
Pe	Péclet Number
Da	Damköhler number
ABP	Active brownian particle
EHD	Electrohydrodynamics
EKI	Electrokinetic
EKI	Electro-elastic instability
FD	Fractional derivative
FPDE	Fractional partial differential equation
RL	Riemann–Liouville

Chapter 1
Introduction

Abstract This chapter introduces complex fluids and specifically, polymeric fluids, and delves into the many strategies used to measure and model them. The motivation for considering sub-diffusive flows is introduced, along with an introduction to diffusive motions and anomalous diffusion.

1.1 Introduction to Complex Fluids

The field of complex fluids, or the study of the wide range of fluids and fluid-like materials that aren't simply homogeneous, memoryless and isotropic, is vast. There are a myriad of ways in which real-world systems can deviate from the standard Newtonian fluid. To give just a few examples:

Suspensions and colloids have inclusions that behave differently from the fluid around them. If the inclusions are solid, we call the system a suspension; if they are liquid, it is an emulsion. The size of the inclusions defines whether or not the material is a colloid, a term that applies when the particles or droplets are small enough to be strongly affected by Brownian motion (typically sub-micron sized). Examples of colloids include emulsions such as milk (which has suspended fat droplets) [1] and suspensions such as ink (which has suspended solid pigment) [2]. Larger inclusions are less strongly affected by Brownian motion; examples here include some muds [3] but also material with more obvious *bits*, such as concrete [4] and even lava [5], which can have solid rocks transported along with molten material.

Granular media consists of individual solid particles, which can stack as a solid or flow in a liquid-like continuum manner. Examples include grain [6], sand [7], pebbles and even snow [8] in avalanche conditions.

Gels arise in food [9] and healthcare [10] products, as well as widely in biology [11]. The underlying molecular structure is networks that span the material, making it look like a solid, but fluid can flow through the spaces within the network [12] and in a weak gel, the bonds can be broken by stresses, allowing the material to truly flow [13].

H. Wilson et al., *Viscoelastic Subdiffusive Flows*, Fluid Mechanics and Its Applications 138, https://doi.org/10.1007/978-981-96-0135-6_1

Active matter is a broad term used to describe any particle that can move without external forcing [14]. Obvious examples come from biology—swimming bacteria [15], flocking birds [16], and growing cells [17]—but this behaviour can also be created artificially by chemical means, using inventions such as Janus particles.

Polymers are long or large molecules that are large enough that their shape can be deformed by flow. Examples include natural materials like rubber and cellulose, but also materials created to serve an industrial purpose like coatings, adhesives and fibres [18]. Either in solution or in melt form, they exhibit viscoelastic properties under flow [19].

It is clear from the diversity of materials and microstructures here that to model all of these with a single framework would be very challenging; in the next sections, we will see an overview of just one special case, that of polymers, as an introduction to the studies in the main body of the book.

1.2 Introduction to Polymeric Fluids

Polymers are everywhere, from foods [20] to industrial materials such as plastics [21], and even our own DNA [22]. In fluid state (whether dissolved in a solvent liquid or in pure liquid form as a melt), they exhibit a phenomenology that is quite distinct from ordinary viscous fluids.

One of the most common observations of polymeric fluids is that their shear viscosity is not constant; rather, it decreases with increasing flow rate, a phenomenon known as shear-thinning [23]. This generally occurs because of the large molecules becoming aligned with the flow during prolonged shear, allowing them to pass more easily past one another. Another is strain hardening [24], the viscosity as measured in an extensional (stretching) flow increases as the fluid is stretched. This is also a result of the polymers aligning with the flow, but this time, their alignment resists the flow rather than making it easier to flow. And finally, because of their ability to be aligned by flow, these materials are viscoelastic [19]: in aligning the polymers, elastic energy is stored in these molecules, which can be released at a later time, giving the fluid a memory of past deformation. Perhaps the easiest of these phenomena to quantify is viscoelasticity, at least in the case of small deformations. This area is called linear rheology, introduced in the following section.

Linear rheology

The linear rheology of a material is its stress response to deformations sufficiently small in amplitude, such that the microstructure of the material remains intact [25]. It is often measured using small-amplitude oscillatory shear in either a Couette device (two concentric cylinders with a small gap between them) or a cone-and-plate device, in which the upper (cone) surface rotates above a static lower plate. Either of these devices produces a good approximation to simple shear in a thin film of the material sample. For small enough oscillations, a sinusoidal strain profile is guaranteed to produce a sinusoidal stress response with the same frequency. So if we impose a strain $\gamma = \gamma_0 \sin \omega t$, with associated shear rate $\dot{\gamma} = \gamma_0 \omega \cos \omega t$, we can measure the resultant shear stress,

$$\sigma = \gamma_0 \left(G' \sin \omega t + G'' \cos \omega t \right).$$

The two coefficients G' and G'', each of which depends on the frequency ω, are the key components of linear rheology; they are often combined to give the complex modulus $G^* = G' + iG''$.

The storage modulus $G'(\omega)$ captures the elastic response of the material or its ability to store energy. Many complex fluids have the ability to relax stresses over very long timescales, resulting in small values of G' for small frequencies ω, while at short timescales (fast oscillations, large ω), they behave in a more solid-like manner and, G' often increases towards high frequencies. A constant value of G' would indicate solid-like behaviour.

The loss modulus $G''(\omega)$ captures the viscous response of the material or its ability to dissipate energy. For a purely viscous Newtonian fluid with viscosity η, we would have $G'' = \omega\eta$ for all frequencies, but complex fluids usually show some memory effects at high frequency, implying that G'' is reduced.

1.3 Modelling Polymeric Fluids

In creating a constitutive model for a polymeric fluid, there are essentially two possible philosophical approaches. The researcher can observe the phenomenology of their fluid [26, 27] (measuring its linear rheology, for example, or the viscosity in steady shear flow as a function of shear rate) and then fit a mathematical function to that measurement. This is the empirical approach and has a long history in polymeric fluids. The alternative approach is to produce models that try to capture the underlying physics of the polymer dynamics and deduce constitutive models from the physical model. This physics-based approach often results in constitutive laws that are a less good direct fit to experimental observations but from which real understanding can be deduced.

Of course, many models that are used largely empirically have some origin in physics-based modelling, and a physics-based model that does not capture at least qualitatively the true behaviour of some real materials would be discarded even before publication, so the separation into these two categories is rather subjective. The more complex models listed here as empirical are almost all derived at some level from a degree of microscopic modelling.

1.3.1 Empirical Models

There are far too many models available to give a comprehensive overview, indeed; the catalogue of potential constitutive models is a book in itself [28]. Here, we simply give a flavour of some of the most commonly used models. The first group of models is the generalised Newtonian fluids, which set the shear viscosity η as a function of the instantaneous local shear rate $\dot{\gamma}$, aiming to fit observations from steady shear experiments as closely as possible.
The power-law model is the simplest of the generalised Newtonian fluids, having only two parameters to fit:

$$\eta = K\dot{\gamma}^{n-1}. \tag{1.1}$$

Usually $0 < n < 1$, to capture shear-thinning [29], but while the power-law may fit real fluid rheology well over the range of measurable shear rates, the viscosity diverges as $\dot{\gamma} \to 0$ so the simplicity of the model comes at a serious price. The special case $n = 1$ is equivalent to a Newtonian fluid.

The Carreau model [30] allows us to reach both low and high shear rates without the model breaking down; it has four disposable constants to govern the viscosity at high and low shear rates, and the power-law behaviour with which it approximates the behaviour between the two,

$$\eta = \eta_\infty + (\eta_0 - \eta_\infty)\left(1 + (\lambda\dot{\gamma})^2\right)^{(n-1)/2}. \tag{1.2}$$

It reduces to a Newtonian fluid if $n = 1$ or $\eta_0 = \eta_\infty$, but in general we expect $\eta_0 > \eta_\infty$ as most polymer systems are shear-thinning.

The Cross model [31] is in many ways similar to the Carreau model: it has viscosity plateaus at high and low shear rates and a power-law like transition between the two, and it is governed by four physical parameters,

$$\eta = \eta_\infty + \frac{\eta_0 - \eta_\infty}{1 + (k\dot{\gamma})^n}. \tag{1.3}$$

The next group of models introduce viscoelasticity, aiming to fit linear rheology data from oscillatory shear. These naturally begin to create fully tensorial models that are capable of responding to flows other than simple shear (even if they have been fit only to shear data). Here, we will use τ to denote the extra stress due to the presence of the polymer and L to denote the velocity gradient tensor.

A linear viscoelastic fluid can be defined [32] in terms of its response to a step strain, the relaxation function $G(t)$,

$$\tau = \int_{\infty}^{t} G(t - t')\mathsf{L}(t')\,\mathrm{d}t'. \tag{1.4}$$

The scalar function $G(t)$ can be found from oscillatory shear experiments through the relation,

$$G^*(\omega) = G'(\omega) + iG''(\omega) = i\omega \int_0^\infty G(s) \exp[-i\omega s]\, ds, \tag{1.5}$$

which is straightforward to invert as it is a simple Fourier transform.

Some of the most common choices for $G(t)$ are single or multi-mode exponential forms,

$$G(t) = G_0 \exp[-\lambda t] \quad \text{or} \quad G(t) = \sum_{j=1}^{N} G_j \exp[-\lambda_j t]. \tag{1.6}$$

Upper-Convected Maxwell model. James Clerk Maxwell [33] was the first to introduce viscoelasticity as a constitutive model, combining an elastic spring and a viscous dashpot to produce a stress model (in shear) that has both fluid-like and solid-like properties according to the timescale over which it is observed,

$$\sigma + \lambda \frac{d\sigma}{dt} = \eta \dot{\gamma}. \tag{1.7}$$

This has been extended to a fully three-dimensional model appropriate to a range of flows, now known as the Upper-Convected Maxwell model or UCM. The derivative introduced here is the upper-convected derivative, which is to line elements what the material derivative is to point elements in the fluid,

$$\tau + \lambda \overset{\triangledown}{\tau} = 2\eta \mathsf{D}. \tag{1.8}$$

We've introduced the symmetrised rate-of-strain tensor $\mathsf{D} = \frac{1}{2}\left(\mathsf{L} + \mathsf{L}^\top\right)$. This model has constant shear viscosity and shows some of the anisotropic stresses seen in shear flow for polymeric fluids, but it has issues with singularities in strong extensional flows, known as the High Weissenberg Number Problem (HWNP). The UCM model can easily be generalised to the Oldroyd-B model (see details below) by the addition of a solvent viscosity term to the stress tensor as well as the usual isotropic pressure.

The Giesekus model [34] introduces a nonlinear term to the polymer stress,

$$\tau + \lambda \overset{\triangledown}{\tau} + \frac{\alpha\lambda}{\eta} \tau \cdot \tau = 2\eta \mathsf{D}. \tag{1.9}$$

The nonlinear term regularises the problems seen by UCM in extensional flows and also introduces a shear-thinning viscosity.

The Phan-Thien-Tanner model (PTT) [35] is rather similar, and also very widely used; its nonlinear term is slightly different,

$$\tau + \lambda \overset{\triangledown}{\tau} + \left[\exp\left(\frac{\lambda}{\eta} \operatorname{trace}\left(\tau^p\right)\right) - 1\right] \tau^p = 2\eta \mathsf{D}. \tag{1.10}$$

This is the exponential form; the exponential is sometimes expanded to linear order or (much less widely used) quadratic order in its argument to form simpler models which are also known as PTT. Like the Giesekus model, it improves many of the major limitations of the simpler UCM model.

1.3.2 Dumbbell Models of Polymer Solutions

The simplest model to allow both elastic and fluid-like behaviour is produced by a pair of beads linked by an elastic spring. The dumbbell itself can be advected by the flow, but if it becomes stretched by the flow, it will relax back towards its original conformation. There are a number of models built on the foundation of dumbbell models.

The Oldroyd B model. Given a linear spring law, the dumbbell paradigm produces the Oldroyd-B model (first proposed by [36], but the microscopic derivation followed later),

$$\tau + \lambda_1 \overset{\nabla}{\tau} = 2\eta \left(\mathsf{D} + \lambda_2 \overset{\nabla}{\mathsf{D}} \right). \tag{1.11}$$

The Oldroyd-B model has a single relaxation time (λ_1), and the other time constant, λ_2, is known as the retardation time and quantifies the relative effect of the solvent to the polymer. In the special case $\lambda_2 = 0$, it reduces to the UCM model. The model can also be expressed as the sum of a Newtonian solvent part and a single elastic mode (setting $\overline{\eta} = \lambda_2\eta/\lambda_1$ and $\overline{G} = (1 - \lambda_2/\lambda_1)\eta/\lambda_1$),

$$\tau = \overline{G}\mathsf{I} + 2\overline{\eta}\mathsf{D} + \overline{G}\mathcal{C} \qquad \overset{\nabla}{\mathcal{C}} = -\frac{1}{\lambda_1}(\mathcal{C} - \mathsf{I}). \tag{1.12}$$

In this formulation, the tensor $\mathcal{C}$ is known as the conformation tensor, as it captures the polymer shape.

Multiple Maxwell modes. The linear rheology of the Oldroyd-B model is a solvent contribution added to a single decaying exponential, that is, a special case of the multi-mode exponential spectrum given in equation (1.6). So, we can see the Maxwell/Oldroyd constitutive equations as a way to generalise an exponential decay from linear rheology to a full constitutive equation. This leads us naturally to two classical predictions of the linear spectrum.

Rouse model. In the Rouse model [37], multiple beads are connected together by a series of linear springs, and there is no excluded volume interaction between the beads. Each bead moves under the action of Brownian motion, the spring force and a drag force.

For a chain with $N + 1$ beads and N springs, we can represent the motion of each bead k as a sum of normal modes $\cos(jk\pi/N)$. Solving for the relaxation of modes varying from the whole length of the chain $j = 1$ to a single monomer $j = N$ gives relaxation times which scale as $\lambda_j \sim 1/j^2$ and all have the same elastic modulus $G_j = G$ in equation (1.6).

Zimm model. In the Zimm model [38], interactions between different parts of the chain, mediated by the solvent, are included. The elastic modulus of each mode is unchanged, but the dependence of the relaxation time on mode number is different from the Rouse modes. In the Zimm model, we have $\lambda_j \sim 1/j^{3\nu}$ where ν is the excluded volume exponent, a parameter capturing how good the solvent is. For a good solvent (in which polymer chains are able to swell) $\nu \approx 3/5$, while for a poor solvent, it can be much lower; the point $\nu = 1/2$ where excluded volume effects balance against polymer solvent effects is called a Θ-solvent.

FENE models. All purely exponential modes are susceptible to potential singularities in extensional flow: if the rate of extension exceeds the rate of relaxation of the slowest mode, then springs can become stretched with no limit, which is not only unphysical but leads to mathematical difficulties as well. Since the unbounded extension of the dumbbells is unphysical, one way to resolve the problem is to introduce a nonlinear spring law such that the springs can no longer stretch without being bound. The dumbbell models that result from this modification are called Finitely Extensible Nonlinear Elastic (FENE) dumbbells. The standard modification to the spring law is to set,

$$\mathbf{F} = k\mathbf{r}f(|\mathbf{r}|), \qquad f(R) = \frac{1}{1 - R^2/b^2}, \tag{1.13}$$

where $\mathbf{r}$ is the end-to-end vector of the dumbbell, and the new parameter b denotes the maximum possible extension of the dumbbell. Unfortunately, the derivation of a constitutive model from an ensemble of such dumbbells is not possible without making some closure approximation. One such approximation is the preaveraging assumption due to Peterlin [39], which results in the FENE-P model,

$$\tau = Gf(R)\mathcal{C} \qquad \mathcal{C} + \frac{\lambda}{f(R)}\overset{\triangledown}{\mathcal{C}} = \frac{\lambda}{f(R)}\mathsf{I} \qquad R^2 = \mathrm{trace}\,(\mathcal{C})\,. \tag{1.14}$$

Another possibility is the FENE-CR model [40], which is designed to have constant shear viscosity to allow the separation of viscoelastic from shear-thinning effects,

$$\tau = Gf(R)\mathcal{C} \qquad \mathcal{C} + \frac{\lambda}{f(R)}\overset{\triangledown}{\mathcal{C}} = \mathsf{I} \qquad R^2 = \mathrm{trace}\,(\mathcal{C})\,. \tag{1.15}$$

1.3.3 Microscopic Models of Polymer Melts

All the models described above are based on the fundamental assumption that our polymeric fluid is a solution, that is, molecules reasonably well separated within a Newtonian solvent. However, in practice, polymers are often processed in melt form, in which the suspending fluid is effectively a soup of the same molecules. There has been significant progress in recent years in modelling the rheology of polymer melts: here, we give a brief overview of a few of the key concepts and models. A much more thorough survey of recent developments is given by Dealy et al. [41].

1.3.3.1 Reptation

Consider a linear polymer in a melt of other similar polymers. When a macroscopic flow is applied to the whole melt, individual polymers can become stretched along their length. In solution, they would relax this stretch according to the Rouse or Zimm timescales, each mode independent of the others. But now, their lateral motion is restricted by the presence of other polymer chains. The stretch along their length (called chain stretch) can relax in roughly the usual way, perhaps with higher friction than would be seen in a true solvent, but the orientation of the chain is held in place by the polymers around it. The only way it can change direction is to migrate, snake-like, along its length. This motion, which was introduced by de Gennes [42], is known as reptation. The reptation time for long polymers is much longer than the Rouse time, and the material's relaxation dynamics will depend strongly on the molecular weight of the polymer.

While the relaxation spectrum gives us the linear viscoelastic properties of the material, it doesn't predict what happens in larger deformations. In recent years there has been a huge leap forward in the development of full, tensorial constitutive laws that build on reptation and the theories that have followed it. Here, we present just one such model, the Rolie-Poly model [43]. This captures the dynamics of linear polymers, including reptation, but also an effect called convective constraint release (CCR), which accounts for the reduction in constraints on a given polymer because of the reptation (not just the chain retraction) of the polymers surrounding it.

$$\overset{\nabla}{\tau} = -\frac{1}{\lambda_d}(\tau - \mathsf{I}) - \frac{2(1-\sqrt{3/\mathrm{tr}})}{\lambda_r}\left(\tau + \beta\left(\tfrac{\mathrm{tr}}{3}\right)^{\delta}(\tau - \mathsf{I})\right), \qquad \mathrm{tr} = \mathrm{trace}(\tau). \tag{1.16}$$

The constitutive equation is rather unwieldy for analytical use, but still far simpler than the full physical model on which it is based.

1.3.3.2 Branched Polymers

If the polymers in the melt are not linear, i.e. their underlying molecular structure has branches, then the spectrum of relaxation times can be even longer as the branchpoints prevent simple reptation motions. Instead, any stress built up in the central backbone can only relax by the slow diffusive motion of the branchpoints, on a timescale on which all the other chains attached to each branchpoint have already relaxed their stress and orientation. This concept results in the pompom model [44] (so-called because of the shape of the collection of dangling arms at the end of each branchpoint),

$$\frac{D\varepsilon}{Dt} = \varepsilon \mathsf{L} : \mathsf{S} - \frac{1}{\lambda_s}(\varepsilon - 1), \qquad \frac{Ds_c}{Dt} = \left(\frac{qs_b}{2} + s_c\right) \mathsf{L} : \mathsf{S} - \frac{1}{2\lambda_a(s_c)},$$
$$\tau = \frac{15}{4} G_0 \phi_b \left(\phi_b \varepsilon^2(t) + \frac{2qs_c(t)}{2qs_a + s_b} \right) \mathsf{S}(t), \tag{1.17}$$
$$\mathsf{S} = \mathcal{C} / \operatorname{trace}(\mathcal{C}), \qquad \overset{\nabla}{\mathcal{C}} = -\frac{1}{\lambda_b}(\mathcal{C} - \mathsf{I}).$$

Here s_a is the scaled molecular weight of the arms, and s_b of the backbone, and q is the number of arms at each branchpoint. The new variables here are s_c, which measures the withdrawal and alignment of the branches, ε, which captures the stretch of the backbone, and S, which is the orientation tensor (scaled conformation tensor) for the central backbone. The timescales are λ_s, the stretch relaxation time, λ_b, the reptation time, which depends on s_c, and $\lambda_a(s)$, the relevant relaxation time at distance s from the branchpoint.

As we can see, with more physics incorporated into each model, the governing equations become more complex. More recent work includes branch-on-branch (BoB) calculations [45] of the relaxation spectra for very complex shaped molecules and much more.

1.4 Rheology and Diffusion

Traditionally, the properties of complex fluids have been investigated using lab-scale rheological measurements such as oscillatory shear (as above) or shear startup or, latterly, extensional flows via either a capillary breakup rheometer [46] or a filament-stretching rheometer [47, 48]. The field of *microrheology* is a more recent invention, following a breakthrough by Mason and Weitz [49], who showed that the motion of a small particle dispersed within a complex fluid, moving under thermal fluctuations, provides a close representation of the linear rheology of the suspending fluid.

1.4.1 Classical Diffusion

Diffusion was first observed by a botanist, Robert Brown [50], watching pollen grains under a microscope and deducing that they were being knocked by air molecules. In his honour, random motion following certain set mathematical rules is now known as *Brownian motion*. In its simplest form, we can consider one-dimensional Brownian motion as a continuum limit of the motion of a random walk. A particle which takes a step Δx or $-\Delta x$ at random, in each timestep Δt, undertakes a *one-dimensional random walk*, also known as a *drunkard's walk*. At time $2m\Delta t$, its position can be described by a binomial probability distribution with zero mean and a variance which scales with m.

Moving to the continuum limit, the limit when $m \to \infty$, while $t \equiv 2m\Delta t$ remains finite, we need to scale $\Delta x = (2D\Delta t)^{1/2}$ for a fixed finite parameter D in order to get a convergent value for the probability density, $\rho(x,t)\,\mathrm{d}x = \mathbb{P}(x,t)$. With that one distinguished scaling, we find that the probability density function satisfies the one-dimensional heat equation,

$$\frac{\partial \rho}{\partial t} = D\frac{\partial^2 \rho}{\partial x^2}. \tag{1.18}$$

This generalises naturally to three dimensions; if a collection of Robert Brown's pollen grains has spatial density $\rho_0(\mathbf{x})$, then, under the action of the air molecules jiggling them around, that density will evolve according to the heat equation,

$$\frac{\partial \rho}{\partial t} = D\nabla^2\rho. \tag{1.19}$$

The heat equation appears in a huge range of physical phenomena, from the transport of heat through a homogeneous solid to the diffusion of pollen particles in still air. In fluid flows, it is typically extended to the *advection-diffusion equation*,

$$\frac{\partial \rho}{\partial t} + \mathbf{v}\cdot\nabla\rho = D\nabla^2\rho, \tag{1.20}$$

where $\mathbf{v}$ is the velocity of the fluid in which the density is evolving.

1.4.2 Anomalous Diffusion

The key property of classical diffusion is that the root mean square distance travelled by a diffusing particle scales with the square root of time:

$$\langle x^2\rangle \sim t \qquad \langle x^2\rangle^{1/2} \sim t^{1/2}. \tag{1.21}$$

In this book, we will investigate some of the phenomena that arise when the diffusive motion does not fit the classical scaling. In general, the term *anomalous diffusion* is used to describe motions that are random but whose average displacement obeys a different, non-classical scaling:

$$\langle x^2\rangle^{1/2} \sim t^{\alpha}. \tag{1.22}$$

These motions are known as *sub-diffusive* if $\alpha < 1/2$, and *super-diffusive* if $\alpha > 1/2$.

Fractional heat equation. The fractional heat equation is related to the standard heat equation in a similar way that anomalous diffusion is related to the classical diffusion, which is outlined as follows: the standard n-dimensional Laplacian operator,

$$\Delta\rho \equiv \nabla^2\rho = \sum_{i=1}^{n} \frac{\partial^2\rho}{\partial x_i^2}, \tag{1.23}$$

must first be generalised to fractional powers. There are a plethora of potential definitions of the fractional Laplacian [51], all of which are equivalent to one another in the function spaces in which they co-exist. Perhaps the easiest to understand is that based on the Fourier transform. This transform is defined as,

$$\mathcal{F}f(\mathbf{y}) = \int_{\mathbb{R}^n} f(\mathbf{x})e^{-i\mathbf{y}\cdot\mathbf{x}}\,\mathrm{d}^n\mathbf{x}. \tag{1.24}$$

For the classical Laplacian,

$$\mathcal{F}(\Delta f)(\mathbf{y}) = -|\mathbf{y}|^2\mathcal{F}f(\mathbf{y}), \tag{1.25}$$

so, by analogy, one can define the fractional Laplacian via the condition,

$$\mathcal{F}(-(-\Delta)^\alpha f)(\mathbf{y}) = -|\mathbf{y}|^{2\alpha}\mathcal{F}f(\mathbf{y}). \tag{1.26}$$

This allows us to write the fractional heat equation,

$$\frac{\partial\rho}{\partial t} + (-\Delta)^{2\alpha}\rho = 0,$$

analogous to the classical diffusion equation (1.19).

1.4.3 *Relation Between Rheology and Diffusion*

Suppose a particle is diffusing under thermal forcing through a complex medium. Mason and Weitz [49] demonstrated a relation between the diffusion of the particle and the relaxation spectrum $G(t)$ of the medium in which it is diffusing. They measure the mean square displacement of the particle, $\langle\Delta r^2(t)\rangle$ as a function of time, and find its Laplace transform $\langle\Delta\tilde{r}^2(s)\rangle$ using the definition,

$$\tilde{f}(s) = \int_0^\infty f(t)e^{-st}\,\mathrm{d}t. \tag{1.27}$$

They were able to show that, to a good approximation, the Laplace transform of the linear spectrum of the suspending fluid is given by,

$$\tilde{G}(s) = \frac{s}{6\pi a}\left[\frac{6k_B T}{s^2\langle \Delta\tilde{r}^2(s)\rangle} - ms\right], \tag{1.28}$$

in which the test particle has radius a and mass m, k_B is the Boltzmann constant and T the temperature. The last term is usually negligible, so we can see that for classical diffusion in which $\langle \Delta\tilde{r}^2(s)\rangle = 6D/s^2$ we recover $\tilde{G}(s) = k_B T s/(6\pi a D)$ which yields a constant viscosity and the standard Stokes–Einstein relation $\eta = k_B T/(6\pi a D)$.

If, on the other hand, the relaxation spectrum is not simply a Newtonian fluid, equation (1.28) allows us to predict the rheology of the fluid from observations of the diffusion of the particle [52]. A fluid with a Rouse relaxation spectrum, for example, will produce anomalous diffusion.

1.4.4 *Relevance of Anomalous Diffusion*

Anomalous diffusion can occur in a wide range of different physical scenarios, including the presence of walls (which can hinder diffusion) or active matter (which can enhance it). A recent review by Oliveira [53] lists many references which have focused on anomalous diffusion in complex fluids, including [54–58]. It is clear that anomalous diffusion occurs much more widely than what had previously been realised.

1.5 Conclusions

We have given a brief overview of the ubiquity of complex fluids and a slightly more thorough survey of the history and current state of the art in the modelling of polymeric fluids. Of course, except in the very simplest flow situations, these constitutive models have no use without numerical simulations to accompany them.

In this chapter, we have seen that there is a fundamental relation between diffusion and rheology, with a wide range of examples in complex fluids, including some classical models of polymer melts producing anomalous diffusion. It is clear that fluids built from a basis of anomalous diffusion provide a very exciting potential framework for further study. In the remainder of this work, we present some explorations of the sub-diffusive regime, including studies of stability and rheodynamical properties.

References

1. Bienvenue A, Jiménez-Flores R, Singh H (2003) Rheological properties of concentrated skim milk: Importance of soluble minerals in the changes in viscosity during storage. J Dairy Sci 86(12):3813–3821
2. Woo K, Jang D, Kim Y, Moon J (2013) Relationship between printability and rheological behavior of ink-jet conductive inks. Ceram Int 39(6):7015–7021
3. Dagois-Bohy S, Hormozi S, Guazzelli É, Pouliquen O (2015) Rheology of dense suspensions of non-colloidal spheres in yield-stress fluids. J Fluid Mech 776:R2
4. Hu J, Wang K (2011) Effect of coarse aggregate characteristics on concrete rheology. Constr Build Mater 25(3):1196–1204
5. Birnbaum J, Lev E, Llewellin EW (2021) Rheology of three-phase suspensions determined via dam-break experiments. Proc R Soc A 477:20210394
6. Fullard LA, Breard ECP, Davies CE, Godfrey AJR, Wade MFA, Dufek J, Lube G (2019) The dynamics of granular flow from a silo with two symmetric openings. Proc R Soc A 475:20180462
7. Hamzah M, Al-Hashemi B, Baghabra Al-Amoudi OS (2018) A review on the angle of repose of granular materials. Powder Technol 330:397–417
8. Nicot F (2004) Constitutive modelling of snow as a cohesive-granular material. Gran Matter 6:47–60
9. Nazir A, Asghar A, Aslam Maan A (2017) Food gels: gelling process and new applications. In: Ahmed J, Ptaszek P, Basu S (eds) Advances in food rheology and its applications, Woodhead publishing series in food science, technology and nutrition. Woodhead Publishing, pp 335–353
10. Fiorillo L, Romano GL (2020) Gels in medicine and surgery: current trends and future perspectives. Gels 6:48
11. Allen A, Hutton DA, Pearson JP, Sellers LA (1984) Mucus glycoprotein structure, gel formation and gastrointestinal mucus function. In: Ciba foundation symposium, vol 109, pp 137–156
12. Duda FP, Souza AC, Fried E (2021) Fluid flow and interface motion in gels: a finite-strain theory and its application to a channel flow problem. J Mech Phys Solids 155:104566
13. Ross-Murphy SB, Shatwell KP (1993) Polysaccharide strong and weak gels. Biorheology 30:217–227
14. Hallatschek O, Datta SS, Drescher K, Dunkel J, Elgeti J, Waclaw B, Wingreen NS (2023) Proliferating active matter. Nat Rev Phys 5:407–419
15. Needleman D, Dogic Z (2017) Active matter at the interface between materials science and cell biology. Nat Rev Mater 2:17048
16. Gompper G et al (2020) The 2020 motile active matter roadmap. Cond Matter 32:193001
17. Morris RG, Rao M (2019) Active morphogenesis of epithelial monolayers. Phys Rev E 100:022413
18. Namazi H (2017) Polymers in our daily life. Bioimpacts 7:73–74
19. Ferry JD (1980) Viscoelastic properties of polymers. Wiley, New York
20. Shah P (2020) Polymers in food. In: Polymer science and innovative applications. Elsevier, pp 567–592
21. Dealy JM, Broadhead TO (1993) Process rheometers for molten plastics: a survey of existing technology. Polym Eng Sci 33:1513–1523
22. Bravo-Anaya LM, Pignon F, Martínez FAS, Rinaudo M (2016) Rheological properties of DNA molecules in solution: molecular weight and entanglement influences. Polymers 8:279
23. Ryder JF, Yeomans JM (2006) Shear thinning in dilute polymer solutions. J Chem Phys 125(19):194906
24. van Melick HGH, Govaert LE, Meijer HEH (2003) On the origin of strain hardening in glassy polymers. Polymer 44(8):2493–2502
25. Barnes HA, Hutton JFJ, Walters K (1989) An introduction to rheology. Elsevier
26. Walters K (1975) Rheometry. Chapman and Hall
27. Dealy JM (1982) Rheometers for molten plastics. Van Nostrand Reinhold

28. Larson RG (1988) Constitutive equations for polymer melts and solutions. Butterworths, Boston, Massachusetts
29. Ostwald W (1929) Über die rechnerische Darstellung des Strukturgebietes der Viskosität. Koll Z 47:176–187
30. Carreau PJ (1972) Rheological equations from molecular network theories. Trans Soc Rheol 16:99–127
31. Cross MM (1965) Rheology of non-Newtonian fluids: a new flow equation for pseudoplastic systems. J Coll Sci 20(5):417–437
32. Tschoegl NW (1989) The phenomenological theory of linear viscoelastic behaviour. Springer
33. Maxwell JC (1867) On the dynamical theory of gases. Philos Trans R Soc Lond 157:49–88
34. Giesekus H (1982) A simple constitutive equation for polymer fluids based on the concept of deformation-dependent tensorial mobility. J Non-Newt Fluid Mech 11(1):69–109
35. Phan-Thien N, Tanner RI (1977) A new constitutive equation derived from network theory. J Non-Newt Fluid Mech 2:353–365
36. Oldroyd JG (1950) On the formulation of rheological equations of state. Proc Roy Soc Lond A200:523–541
37. Rouse PE (1953) A theory of the linear viscoelastic properties of dilute solutions of coiling polymers. J Chem Phys 21:1272–1280
38. Zimm BH (1956) Dynamics of polymer molecules in dilute solution: viscoelasticity, flow birefringence and dielectric loss. J Chem Phys 24:269–278
39. Peterlin A (1961) Einfluß der endlichen Moleküllänge auf die Gradientenabhängigkeit des Staudinger-Index. Makromol Chimie 44:338–346
40. Chilcott MD, Rallison JM (1988) Creeping flow of dilute polymer solutions past cylinders and spheres. J Non-Newt Fluid Mech 29:381–432
41. Dealy JM, Read DJ, Larson RG (2018) Structure and rheology of molten polymers: from structure to flow behavior and back again. Hansser
42. de Gennes PG (2003) Reptation of a polymer chain in the presence of fixed obstacles. J Chem Phys 55(2):572–579
43. Likhtman AE, Graham RS (2003) Simple constitutive equation for linear polymer melts derived from molecular theory: Rolie-Poly equation. J Non-Newt Fluid Mech 114(1):1–12
44. Bishko G, McLeish TCB, Harlen OG, Larson RG (1997) Theoretical molecular rheology of branched polymers in simple and complex flows: the pom-pom model. Phys Rev Lett 79(12):2352–2355
45. Das C, Inkson NJ, Read DJ, Kelmanson MA, McLeish TCB (2006) Computational linear rheology of general branch-on-branch polymers. J Rheol 50(2):207–234
46. Bazilevsky AV, Entov VM, Rozhkov AN (2011) Breakup of a liquid bridge as a method of rheological testing of biological fluids. Fluid Dyn 46:613–622
47. Sridhar T, Tirtaatmadja V, Nguyen D, Gupta R (1991) Measurement of extensional viscosity of polymer solutions. J Non-Newt Fluid Mech 40:271–280
48. Tirtaatmadja V, Sridhar T (1993) A filament stretching device for measurement of extensional viscosity. J Rheol 37:1081–1102
49. Mason TG, Weitz D (1995) Optical measurements of frequency-dependent linear viscoelastic moduli of complex fluids. Phys Rev Lett 74:1250
50. Brown R (1828) A brief account of microscopical observations made in the months of June, July and august 1827, on the particles contained in the pollen of plants; and on the general existence of active molecules in organic and inorganic bodies. Philos Mag 4:161–173
51. Kwaśnicki M (2017) Ten equivalent definitions of the fractional Laplace operator. FCAA 20:7–51
52. Fricks J, Yao L, Elston TC, Forest MG (2009) Time-domain methods for diffusive transport in soft matter. SIAM J Appl Math 69(5):1277–1308
53. Oliveira FA, Ferreira RMS, Lapas LC, Vainstein MH (2019) Anomalous diffusion: a basic mechanism for the evolution of inhomogeneous systems. Front Phys 7:18
54. Grmela M, Öttinger HC (1997) Dynamics and thermodynamics of complex fluids. I. Development of a general formalism. Phys Rev E 56

55. Bakk A, Fossum JO, da Silva GJ, Adland HM, Mikkelsen A, Elgsaeter A (2002) Viscosity and transient electric birefringence study of clay colloidal aggregation. Phys Rev E 65:021407
56. Sehnem AL, Aquino R, Campos AFC, Tourinho FA, Depeyrot J, Neto AMF (2014) Thermodiffusion in positively charged magnetic colloids: influence of the particle diameter. Phys Rev E 89:032308
57. Sehnem AL, Neto AMF, Aquino R, Campos AFC, Tourinho FA, Depeyrot J (2015) Temperature dependence of the Soret coefficient of ionic colloids. Phys Rev E 92:042311
58. Cabreira Gomes R, Ferreira da Silva A, Kouyaté M, Demouchy G, Mériguet G, Aquino R, Dubois E, Nakamae S, Roger M, Depeyrot J, Perzynski R (2018) Thermodiffusion of repulsive charged nanoparticles—the interplay between single-particle and thermelectric contributions. Phys Chem Chem Phys 20:16402–16413

Chapter 2
Preliminaries of Fractional Derivatives

Abstract The fractional differential operators have at least sixteen different kinds of definitions, and not all of them are equivalent. This chapter presents the numerical approximations of the most frequently used fractional derivatives, including the Grünwald Letnikov derivative, the Riemann Liouville derivative and the Caputo derivative and list the operator properties relevant in the modelling of anomalous diffusion.

2.1 Introduction

In the realm of mathematical modelling and computational simulations, the use of fractional calculus has gained significant attention due to its ability to describe and analyze complex phenomena that classical integer-order calculus fails to capture adequately. Due to our specific application, we limit our focus throughout this and subsequent chapters on the development of the fractional Riemann Liouville derivative ($_{RL}D(\cdot)$), the fractional Caputo derivative ($_{C}D(\cdot)$) and the fractional Grünwald Letnikov derivative ($_{GL}D(\cdot)$). The definitions of these fractional derivatives are not equivalent. For example, if $f(t)$ is suitably smooth, i.e. $f \in C^m[a,b]$, then

$$_{RL}D^{\alpha}_{a,t}f(t) =_{GL} D^{\alpha}_{a,t}f(t), \tag{2.1}$$

while the Riemann Liouville and Caputo derivative of $f(t)$ have the following relation,

$$_{RL}D^{\alpha}_{a,t}f(t) =_{C} D^{\alpha}_{a,t}f(t) + \sum_{k=0}^{m-1} \frac{f^{(k)}(a)(t-a)^{k-\alpha}}{\Gamma(k+1-\alpha)}, \tag{2.2}$$

where $m-1 < \alpha < m$, m is a positive integer, $\Gamma(\cdot)$ is the Γ-function (refer Definition 2.1), $f \in C^{m-1}[a,t]$ and $f^{(m)}$ is integrable on $[a,t]$. This chapter is organized as follows. Section 2.2 provides a basic framework of the fractional derivatives as well as the properties of the fractional operators needed to develop our mathematical model. Section 2.3 presents a brief background on the numerical methods for fractional derivatives as well as for fractional ordinary and partial differential equations.

H. Wilson et al., *Viscoelastic Subdiffusive Flows*, Fluid Mechanics and Its Applications 138, https://doi.org/10.1007/978-981-96-0135-6_2

2.2 Grünwald Letnikov, Riemann Liouville and Caputo Derivatives

We begin our exposition by outlining some special functions used in the definition of these derivatives, as follows,

Definition 2.1 The Gamma function $\Gamma(z)$ is a fundamental mathematical function that generalizes the concept of the factorial to non-integer arguments. For $z \in \mathbb{C}$ such that $z_r > 0$, $\Gamma(z)$ is defined as,

$$\Gamma(z) = \int_0^\infty t^{z-1} e^{-t} dt. \tag{2.3}$$

The incomplete Gamma function is defined as,

$$\Gamma^*(z, x) = \int_0^x t^{z-1} e^{-t} dt. \tag{2.4}$$

Definition 2.2 For $z \in \mathbb{C}$, the Mittag Leffler function $E_\alpha(z)$ is defined via the following power series representation,

$$E_\alpha(z) = \sum_{k=0}^{\infty} \frac{z^k}{\Gamma(\alpha k + 1)} \qquad \alpha > 0, \tag{2.5}$$

and the generalized Mittag Leffler function $E_{\alpha,\beta}(z)$ is defined as,

$$E_{\alpha,\beta}(z) = \sum_{k=0}^{\infty} \frac{z^k}{\Gamma(\alpha k + \beta)} \qquad \alpha > 0, \beta > 0. \tag{2.6}$$

Using the above defined special functions, we outline the definitions of the fractional derivatives.

Definition 2.3 The Grünwald Letnikov derivative with order $\alpha > 0$ of the given function $f(t), t \in (a, b)$ is defined as,

$$_{GL}D^\alpha_{a,t} f(t) = \lim_{\substack{h \to 0 \\ Nh = t-a}} h^{-\alpha} \sum_{j=0}^{N} (-1)^j \binom{\alpha}{j} f(t - jh). \tag{2.7}$$

Definition 2.4 The Riemann Liouville derivative with order $\alpha > 0$ of the given function $f(t), t \in (a, b)$ is defined as,

$$_{RL}D^\alpha_{a,t} f(t) = \frac{1}{\Gamma(m - \alpha)} \frac{d^m}{dt^m} \int_a^t (t - s)^{m-\alpha-1} f(s) ds, \tag{2.8}$$

where m is a positive integer satisfying the inequality, $m - 1 \leq \alpha < m$.

An important limitation of the Riemann Liouville derivative is highlighted in the following example. Consider $f(t) = t^{\mu}$, where $\mu > -1$. By definition of the Riemann Liouville derivative,

$$\begin{aligned} {}_{RL}D^{\alpha}_{0,t} t^{\mu} &= \frac{d}{dt}\left({}_{RL}D^{-(1-\alpha)}_{0,t} t^{\mu}\right), \\ &= \frac{d}{dt}\left(\frac{\Gamma(\mu+1)}{\Gamma(2-\alpha+\mu)} t^{1-\alpha+\mu}\right), \\ &= \frac{\Gamma(\mu+1)}{\Gamma(1-\alpha+\mu)} t^{\mu-\alpha}. \end{aligned}$$

The above example shows that if k is a non-zero constant, ${}_{RL}D^{\alpha}_{0,t} k = \dfrac{k}{\Gamma(1-\alpha)} t^{-\alpha}$, which is not zero. This fact led Ochmann and Makarov [1] to use the Riemann Liouville definition with $a = -\infty$ (the lower limit in Definition 2.8) because, on one hand, from physical point of view it is required that the fractional derivative of a constant equals zero and on the other hand formula (2.8) gives zero if $a \to -\infty$. The physical meaning of this step is that the starting time of the sub-diffusive process is set to $-\infty$. In such a case, transient effects cannot be studied. However, taking $a = -\infty$ is the necessary abstraction for the consideration of the steady-state processes, for example, for studying the response of the fractional order dynamical system to a periodic input signal in viscoelastic materials.

On the contrary, the Caputo fractional derivatives are consistent (i. e., the Caputo definition of a derivative does not have the above paradox) and are well-suited for modelling initial value problems. The fundamental benefit of Caputo's method is that the initial conditions for fractional differential equations with Caputo derivatives have the same structure as differential equations for integer order. They contain the limiting values for integer order derivatives at $t = a$.

Definition 2.5 The Caputo derivative with order $\alpha > 0$ of the given function $f(t), t \in (a, b)$ is defined as,

$${}_{C}D^{\alpha}_{a,t} f(t) = \frac{1}{\Gamma(m-\alpha)} \int_a^t (t-s)^{m-\alpha-1} f^{(m)}(s) ds, \tag{2.9}$$

where m is a positive integer satisfying $m - 1 < \alpha \leq m$.

2.2.1 Operator Properties of Fractional Derivatives

Next, we list some properties of fractional derivatives useful for modelling the dynamics of viscoelastic sub-diffusive flows. The proofs for the following propositions can be found in the classical textbooks by Miller [2] and Podlubny [3].

Proposition 2.1 (Commutativity of Riemann Liouville derivative)

The Riemann Liouville fractional derivative operators satisfy the following properties,

$$ {}_{RL}D^{\alpha}_{a,t}D^{-\alpha}_{a,t}f(t) = f(t), \tag{2.10a}$$

$$ D^{-\alpha}_{a,t}\left({}_{RL}D^{\alpha}_{a,t}f(t)\right) = f(t) - \sum_{j=1}^{m}\left[{}_{RL}D^{\alpha-j}_{a,t}f(t)\right]_{t=a}\frac{(t-a)^{\alpha-j}}{\Gamma(\alpha-j+1)}, \tag{2.10b}$$

where $m-1 \le \alpha < m$, m is a positive integer. If $f(t)$ has sufficient number of continuous derivatives, then $\left[{}_{RL}D^{\alpha-j}_{a,t}f(t)\right]_{t=a}=0$, $j=1,2,\cdots m$ is equivalent to $f^{(j)}(a)=0$, $j=0,1,\cdots m-1$.

The above proposition may be generalized as follows,

$$ {}_{RL}D^{\alpha}_{a,t}D^{-\beta}_{a,t}f(t) = {}_{RL}D^{\alpha-\beta}_{a,t}f(t), \tag{2.11a}$$

$$ D^{-\beta}_{a,t}\left({}_{RL}D^{\alpha}_{a,t}f(t)\right) = {}_{RL}D^{\alpha-\beta}_{a,t}f(t) - \sum_{j=1}^{m}\left[{}_{RL}D^{\alpha-j}_{a,t}f(t)\right]_{t=a}\frac{(t-a)^{\beta-j}}{\Gamma(1+\beta-j)}, \tag{2.11b}$$

where $m-1 \le \alpha < m, n-1 \le \beta < n$, m, n are positive integers.

Proposition 2.2 (Composition property of Riemann Liouville derivative)

If $m-1 \le \alpha < m, n-1 \le \beta < n$, where m, n are positive integers, ${}_{RL}D^{\alpha+\beta}_{a,t}f(t)$ and ${}_{RL}D^{\alpha}_{a,t}\left({}_{RL}D^{\beta}_{a,t}f(t)\right)$ exist, then,

$$ {}_{RL}D^{\alpha}_{a,t}\left({}_{RL}D^{\beta}_{a,t}f(t)\right) = {}_{RL}D^{\alpha+\beta}_{a,t}f(t) - \sum_{j=1}^{n}\left[{}_{RL}D^{\beta-j}_{a,t}f(t)\right]_{t=a}\frac{(t-a)^{-\alpha-j}}{\Gamma(1-\alpha-j)}. \tag{2.12}$$

Proposition 2.3 (Composition property of Caputo derivative: I)

If $\alpha > 0$, $n-1 < \beta < n$, n is a positive integer, $f \in C^n[a,b]$, then,

$$ D^{-\alpha}_{a,t}\left({}_{C}D^{\beta}_{a,t}f(t)\right) = {}_{C}D^{\beta-\alpha}_{a,t}f(t), \tag{2.13}$$

where $D^{-\alpha}_{a,t}(\cdot)$ is the Riemann Liouville Integral and ${}_{C}D^{\beta-\alpha}_{a,t} = D^{\beta-\alpha}_{a,t}$, if $\beta < \alpha$. Especially,

$$ D^{-\beta}_{a,t}\left({}_{C}D^{\beta}_{a,t}f(t)\right) = f(t) - \sum_{k=0}^{n-1}\frac{f^{(k)}(a)}{\Gamma(k+1)}(t-a)^k. \tag{2.14}$$

Proposition 2.4 (Composition property of Caputo derivative: II)

If $\alpha, \beta > 0$, $f(t)$ is sufficiently smooth, ${}_{C}D^{\beta}_{a,t}D^{-\alpha}_{a,t}f(t)$ and ${}_{C}D^{\beta-\alpha}_{a,t}f(t)$ exists, then,

$$
{}_C D_{a,t}^{\beta}\left(D_{a,t}^{-\alpha} f(t)\right) = \begin{cases} D_{a,t}^{(\alpha-\beta)} f(t), \quad \beta \leq \alpha, \alpha < \beta, \alpha \in \mathbb{N} \\ {}_C D_{a,t}^{\beta-\alpha} f(t) + \sum_{k=0}^{n-m} \dfrac{f^{(k)}(a)}{\Gamma(k+1+\alpha-\beta)}(t-a)^{k+\alpha-\beta}, \\ \qquad \alpha < \beta, m{-}1 < \alpha < m, n{-}1{<}\beta{<}n, m, n{\in}\mathbb{N}. \end{cases} \tag{2.15}
$$

Proposition 2.5 (Composition property of Caputo derivative: III)
If $\beta > 0$, $n - 1 < \beta < n$ and m be any non-negative integer, then,

$$
{}_C D_{a,t}^{\beta}\left({}_C D_{a,t}^{m} f(t)\right) = {}_C\, D_{a,t}^{m+\beta} f(t), \tag{2.16a}
$$

$$
{}_C D_{a,t}^{m}\left({}_C D_{a,t}^{\beta} f(t)\right) = {}_C\, D_{a,t}^{m+\beta} f(t) + \sum_{j=n}^{m+n-1} \frac{f^{(j)}(a)}{\Gamma(1+j-m-\beta)}(t-a)^{j-m-\beta}. \tag{2.16b}
$$

Proposition 2.6 (Linearity)
Fractional integrals and derivatives are linear operators, i.e.,

$$
D^{\alpha}(\lambda_1 f(t) + \lambda_2 g(t)) = \lambda_1 D^{\alpha} f(t) + \lambda_2 D^{\alpha} g(t), \tag{2.17}
$$

where D^{α} represents any fractional derivative or integral.

Proposition 2.7 (Leibniz Rule for Fractional Derivatives for product of functions)
Let $f(t)$ and $g(t)$ be any two functions with derivatives up to n. Then, the Leibniz rule for fractional derivatives is given by,

$$
{}_{RL} D_{a,t}^{\alpha}(g(t) f(t)) = \sum_{k=0}^{n} \binom{\alpha}{k} g^{(k)}(t) f^{(\alpha-k)}(t) - R_n^{\alpha}(t), \tag{2.18}
$$

where $n \geq \alpha + 1$, $f^{(\alpha-k)}(t) = {}_{RL}\, D_{a,t}^{\alpha-k} f(t)$, and the remainder $R_n^{\alpha}(t)$ is given by,

$$
R_n^{\alpha}(t) = \frac{(-1)^n (t-a)^{n-\alpha+1}}{n!\,\Gamma(-\alpha)} \int_0^1 F_a(t, \zeta, \eta) d\zeta d\eta, \tag{2.19}
$$

where

$$
F_a(t, \zeta, \eta) = f(a + \eta(t-a)) g^{(n+1)}(a + (t-a)(\zeta + \eta - \zeta\eta)).
$$

Proposition 2.8 (Laplace Transform)
The Laplace transform of the Riemann Liouville derivative operator with order α, $m - 1 \leq \alpha < m$ is given by,

$$
L\{{}_{RL} D_{0,t}^{\alpha} f(t); s\} = s^{\alpha} L\{f(T); s\} - \sum_{k=0}^{m-1} s^k \left[{}_{RL} D_{0,t}^{\alpha-k-1} f(t)\right]_{t=0}. \tag{2.20}
$$

The Laplace transform of the Caputo derivative operator with order α, $m-1<\alpha\leq m$, *is given by*

$$L\{_C D_{0,t}^{\alpha} f(t); s\} = s^{\alpha} L\{f(T); s\} - \sum_{k=0}^{m-1} s^{\alpha-k-1} f^{(k)}(0). \tag{2.21}$$

2.3 Numerical Approximations of Fractional Derivatives

Since the Grünwald-Letnikov derivative and the Riemann Liouville derivative are equivalent [4], we mainly consider the numerical methods for the left Reimann Liouville derivative with fractional order $0 < \alpha < 1$ and $1 < \alpha < 2$, which has special importance in the modelling of anomalous diffusion [5].

- **Grünwald-Letnikov approximation**
  If $f(t)$ is suitably smooth, the Grünwald-Letnikov derivative is equivalent to the Reimann Liouville derivative. Denoting by $\omega_j^{\alpha} = (-1)^j \binom{\alpha}{j}$, one gets,

$$\left[{}_{RL}D_{0,t}^{\alpha} f(t)\right]_{t=t_n} \approx \frac{1}{\Delta t^{\alpha}} \sum_{j=0}^{n} \omega_j^{(\alpha)} f(t_{n-j}). \tag{2.22}$$

  Formula (2.22) is convergent for order 1 and for any order $\alpha > 0$, but it may contribute to unstable numerical schemes for $1 < \alpha < 2$ [6]. The shifted Grünwald-Letnikov formula (p shifts, $p \in \mathbb{N}$) is useful for constructing stable numerical schemes and is first-order accurate. It is given by

$$\left[{}_{RL}D_{0,t}^{\alpha} f(t)\right]_{t=t_n} \approx \frac{1}{\Delta t^{\alpha}} \sum_{j=0}^{n+p} \omega_j^{(\alpha)} f(t_{n-j+p}). \tag{2.23}$$

- **L1 method**
  The classical L1 method leads to an unconditionally stable algorithm for the case $0 < \alpha < 1$. Since the relation between the Riemann-Liouville and Caputo derivative is as follows,

$${}_{RL}D_{0,t}^{\alpha} f(t) = {}_C D_{0,t}^{\alpha} f(t) + \frac{f(0)}{\Gamma(1-\alpha)} t^{-\alpha}, \tag{2.24}$$

  the L1 approximation is given by,

$$\left[{}_{RL}D_{0,t}^{\alpha} f(t)\right]_{t=t_n} \approx \frac{f(0)t_n^{-\alpha}}{\Gamma(1-\alpha)} + \frac{1}{\Gamma(1-\alpha)} \sum_{k=0}^{n-1} \int_{t_k}^{t_{k+1}} (t_n - s)^{-\alpha} \frac{f(t_{k+1}) - f(t_k)}{\Delta t} ds,$$

$$= \frac{f(0)t_n^{-\alpha}}{\Gamma(1-\alpha)} + \sum_{k=0}^{n-1} b_{n-k-1}\left[f(t_{k+1}) - f(t_k)\right], \tag{2.25}$$

where $t_0 = 0$ and $b_k = \dfrac{\Delta t^{-\alpha}}{\Gamma(2-\alpha)}\left[(k+1)^{1-\alpha} - k^{1-\alpha}\right]$. Formula (2.25) has the following error estimate,

$$\left| \frac{f(0)t_n^{-\alpha}}{\Gamma(1-\alpha)} + \sum_{k=0}^{n-1} b_{n-k-1}\left[f(t_{k+1}) - f(t_k)\right] - \left[{}_{RL}D_{0,t}^{\alpha} f(t)\right]_{t=t_n} \right| \le C\Delta t^{2-\alpha}, \tag{2.26}$$

where C is a positive constant only dependent on α and f.

- **L2 and L2C method**

  The L2 method for the Reimann Liouville derivative of order α $(1 < \alpha < 2)$, is obtained in a way similar to the L1 method,

$$\begin{aligned}\left[{}_{RL}D_{0,t}^{\alpha} f(t)\right]_{t=t_n} &\approx \frac{f(0)t_n^{-\alpha}}{\Gamma(1-\alpha)} + \frac{1}{\Gamma(2-\alpha)} \\ &\sum_{k=0}^{n-1} \int_{t_k}^{t_{k+1}} s^{1-\alpha} \frac{f(t_n - t_{k+1}) - 2f(t_n - t_k) + f(t_n - t_{k-1})}{\Delta t^2} ds, \\ &= \frac{f(0)t_n^{-\alpha}}{\Gamma(1-\alpha)} + \frac{f'(0)t_n^{1-\alpha}}{\Gamma(2-\alpha)} + \sum_{k=-1}^{n} W_k f(t_{n-k}),\end{aligned} \tag{2.27}$$

where the weights W_k are given by,

$$W_k = \frac{\Delta t^{-\alpha}}{\Gamma(3-\alpha)} \begin{cases} 1, & k=-1, \\ 2^{2-\alpha} - 3, & k = 0, \\ (k+2)^{2-\alpha} - 3(k+1)^{2-\alpha} + 3k^{2-\alpha} - (k-1)^{2-\alpha}, & 1 \le k \le n-2, \\ -2n^{2-\alpha} + 3(n-1)^{2-\alpha} - (n-2)^{2-\alpha}, & k = n-1, \\ n^{2-\alpha} - (n-1)^{2-\alpha}, & k = n. \end{cases}$$

Another variant, the L2C method, is given by,

$$\left[{}_{RL}D_{0,t}^{\alpha} f(t)\right]_{t=t_n} \approx \frac{f(0)t_n^{-\alpha}}{\Gamma(1-\alpha)} + \frac{f'(0)t_n^{1-\alpha}}{\Gamma(2-\alpha)} + \sum_{k=-1}^{n+1} \hat{W}_k f(t_{n-k}), \tag{2.28}$$

where the weights $\hat{W}_k$ are given by,

$$\hat{W}_k = \frac{\Delta t^{-\alpha}}{2\Gamma(3-\alpha)} \begin{cases} 1, & k=-1, \\ 2^{2-\alpha}-2, & k=0, \\ 3^{2-\alpha}-2^{2-\alpha}, & k=1, \\ (k+2)^{2-\alpha}-2(k+1)^{2-\alpha}+2(k-1)^{2-\alpha}-(k-2)^{2-\alpha}, & 2 \le k \le n-2, \\ -n^{2-\alpha}-3(n-1)^{2-\alpha}+2(n-2)^{2-\alpha}, & k=n-1, \\ -n^{2-\alpha}+2(n-1)^{2-\alpha}-(n-2)^{2-\alpha}, & k=n, \\ n^{2-\alpha}-(n-1)^{2-\alpha}, & k=n+1. \end{cases}$$

Both the L2 and the L2C methods are applicable for derivatives of order $1 < \alpha < 2$ and converges with order $O\left(\Delta t^{3-\alpha}\right)$ [4].

2.3.1 *Fractional Ordinary Differential Equations*

In this section, we focus on introducing numerical methods for fractional ordinary differential equations. Since the Reimann-Liouville derivatives can be changed into Caputo ones (see Eq. (2.24)), we study only the Caputo-type ordinary differential equations. Consider the initial value problem,

$$\begin{cases} {}_C D^{\alpha}_{0,t} u(t) = f(t, u(t)), & m-1 < \alpha < m \in \mathbb{Z}^+ \\ u^{(j)}(0) = u_0^j, & j = 0, 1, \cdots, m-1. \end{cases} \tag{2.29}$$

The following two theorems found in [7] aid in the existence and uniqueness of the initial value problem (2.29).

Theorem 2.1 *Assume that* $\mathcal{D} := [0, \chi^*] \times \left[u_0^0 - \delta, u_0^0 + \delta\right]$, *with some* $\chi^* > 0$, *and some* $\delta > 0$, *and let the function* $f : \mathcal{D} \to \mathbb{R}$ *be continuous. Furthermore, define* $\chi := min\{\chi^*, \left(\delta\Gamma(\alpha+1)\ \|f\|_\infty\right)^{1/\alpha}\}$. *Then, there exists a function* $u : [0, \chi] \to \mathbb{R}$ *solving the initial value problem* (2.29).

Theorem 2.2 *Assume that* $\mathcal{D} := [0, \chi^*] \times \left[u_0^0 - \delta, u_0^0 + \delta\right]$, *with some* $\chi^* > 0$, *and some* $\delta > 0$. *Furthermore, let the function* $f : \mathcal{D} \to \mathbb{R}$ *be bounded on* $\mathcal{D}$ *and fulfil a Lipschitz condition with respect to the second variable, i.e.,*

$$|f(t, x) - f(t, y)| \le L\,|x - y|, \tag{2.30}$$

with constant $L > 0$ *independent of* t, x *and* y. *Then denoting* χ *as in theorem 2.1, there exists at most one function* $u : [0, \chi] \to \mathbb{R}$ *solving the initial value problem* (2.29).

Next, we list some numerical methods for Eq. (2.29), without giving the stability and convergence analysis:

- **L1 methods**
For $0 < \alpha < 1$, L1 scheme (2.25) is used to discretize the Caputo derivative in Eq. (2.29), given by:

$$\sum_{j=0}^{n-1} b_{n-j-1}\left(u_{j+1} - u_j\right) = f(t_n, u_n), \tag{2.31}$$

where $u_n \approx u(t_n)$, and $b_j = \frac{\Delta t^{-\alpha}}{\Gamma(2-\alpha)}\left[(j+1)^{1-\alpha} - j^{1-\alpha}\right]$.

- **Grünwald Letnikov formula**
Using the Grünwald Letnikov approximation (2.22), one arrives at the following discretized version of Eq. (2.29)

$$\frac{1}{\Delta t^{\alpha}} \sum_{j=0}^{n} \omega_{n-j}^{(\alpha)} \left[u_j - \sum_{k=0}^{m} \frac{u^k(0)}{k!} t_j^k\right] = f(t_n, u_n), \quad m - 1 < \alpha < m. \tag{2.32}$$

For better numerical stability, one may resort to the shifted Grünwald Letnikov formula (2.23) to approximate the Caputo derivative.

2.3.2 *Fractional Partial Differential Equations*

In this section, we discuss the numerical methods for Caputo-type subdiffusion equation. Let $U(x, t)$ be a suitable smooth function such that it satisfies the time-fractional diffusion equation given by,

$$\begin{aligned} &{}_C D_{0,T}^{\alpha} U = K_\alpha \partial_x^2 U + g(x,t), \qquad (x,t) \in (a,b) \times (0,T], \\ &U(x,0) = \phi_0(x), \qquad x \in (a,b), \\ &U(a,t) = U_a(t), U(b,t) = U_b(t), \qquad t \in (0,T]. \end{aligned} \tag{2.33}$$

The commonly used numerical methods for equation (2.33) are listed as follows:

- **Explicit Euler Type Method (Grünwald Letnikov Approximation)**
Let the Caputo derivative be discretized using the Grünwald-Letnikov approximation (2.22), and the spatial derivative be discretized using the central difference scheme. Let u_i^n denote the approximate value of $U(x_i, t_n)$ where $\Delta x = \frac{b-a}{N}$, $\Delta t = \frac{T}{n_T}$ and $x_i = a + i\Delta x$, $t_n = n\Delta t$. Then,

$$\frac{1}{\Delta t^{\alpha}} \sum_{k=0}^{n} \omega_{n-k}^{(\alpha)} \left(u_i^k - u_i^0\right) = K_\alpha \left(\frac{u_{i+1}^{n-1} - 2u_i^{n-1} + u_{i-1}^{n-1}}{(\Delta x)^2}\right) + g_i^n, \quad i = 1, 2, \cdots, N-1,$$

$$
\begin{aligned}
u_i^0 =& \phi_0(x_i), \quad i = 0, 1, 2, \cdots, N, \\
u_0^n =& U_a(t_n), u_N^n = U_b(t_n),
\end{aligned} \tag{2.34}
$$

where $\omega_k^{(\alpha)} = (-1)^k \binom{\alpha}{k}$. This method is conditionally stable if $\frac{K_\alpha (\Delta t)^\alpha}{(\Delta x)^2} \leq \frac{\alpha}{2}$.

- **Explicit Euler Type Method (L1 Approximation)**
  When Eq. (2.33) is discretized using the L1 method, it reduces to the following,

$$
\begin{aligned}
\frac{1}{\Delta t^\alpha} \sum_{k=0}^{n-1} b_{n-k-1}^{(\alpha)} (u_i^{k+1} - u_i^k) = K_\alpha \left(\frac{u_{i+1}^{n-1} - 2u_i^{n-1} + u_{i-1}^{n-1}}{(\Delta x)^2} \right) + g_i^n, \quad i = 1, 2, \cdots, N-1, \\
u_i^0 = \phi_0(x_i), \quad i = 0, 1, 2, \cdots, N, \\
u_0^n = U_a(t_n), u_N^n = U_b(t_n),
\end{aligned} \tag{2.35}
$$

where $b_k^{(\alpha)} = \frac{1}{\Gamma(2-\alpha)} \left((k+1)^{1-\alpha} - k^{1-\alpha} \right)$. This method is conditionally stable if $\frac{K_\alpha (\Delta t)^\alpha}{(\Delta x)^2} \leq \frac{1 - 2^{-\alpha}}{\Gamma(2-\alpha)}$ [4].

- **Implicit Euler Type Method (Grünwald Letnikov Approximation)**
  The implicit scheme using the Grünwald Letnikov approximation is given by,

$$
\begin{aligned}
\frac{1}{\Delta t^\alpha} \sum_{k=0}^{n} \omega_{n-k}^{(\alpha)} (u_i^k - u_i^0) = K_\alpha \left(\frac{u_{i+1}^n - 2u_i^n + u_{i-1}^n}{(\Delta x)^2} \right) + g_i^n, \quad i = 1, 2, \cdots, N-1, \\
u_i^0 = \phi_0(x_i), \quad i = 0, 1, 2, \cdots, N, \\
u_0^n = U_a(t_n), u_N^n = U_b(t_n),
\end{aligned} \tag{2.36}
$$

where $\omega_k^{(\alpha)} = (-1)^k \binom{\alpha}{k}$. This method is unconditionally stable and is convergent of order $O(\Delta t + \Delta x^2)$.

- **Implicit Euler Type Method (L1 Approximation)**
  The implicit scheme using the L1 method is given by,

$$
\begin{aligned}
\frac{1}{\Delta t^\alpha} \sum_{k=0}^{n-1} b_{n-k-1}^{(\alpha)} (u_i^{k+1} - u_i^k) = K_\alpha \left(\frac{u_{i+1}^n - 2u_i^n + u_{i-1}^n}{(\Delta x)^2} \right) + g_i^n, \quad i = 1, 2, \cdots, N-1, \\
u_i^0 = \phi_0(x_i), \quad i = 0, 1, 2, \cdots, N, \\
u_0^n = U_a(t_n), u_N^n = U_b(t_n),
\end{aligned} \tag{2.37}
$$

where $b_k^{(\alpha)} = \frac{1}{\Gamma(2-\alpha)} \left((k+1)^{1-\alpha} - k^{1-\alpha} \right)$.
This method is unconditionally stable and is convergent of order $O(\Delta t^{2-\alpha} + \Delta x^2)$.

For further exploration of these methods, especially to study the stability and the order of convergence, the readers are requested to refer to the book by Li and Zeng [4].

2.4 Conclusions

This chapter addresses three different kinds of fractional derivatives, relevant in modeling of anomalous diffusion. They are not all mutually equivalent. Among these definitions, the Riemann-Liouville derivative and the Caputo derivative, which are defined on the basis of fractional integral are most frequently used. Stochastic experts, pure mathematicians and physicists would rather use the former, while applied mathematicians and engineers prefer to utilize the latter, mainly due to their respective research backgrounds [8]. We conclude with a remark that fractional calculus is not the mathematical generalization of classical calculus. For example, for Caputo derivative if we fix t and let the order $\alpha \in (n-1, n)$ vary, we have

$$\lim_{\alpha \to (n-1)^+} {}_C D_{0,t}^{\alpha} x(t) = x^{(n-1)}(t) - x^{(n-1)}(0), \quad \lim_{\alpha \to n^-} {}_C D_{0,t}^{\alpha} x(t) = x^{(n)}(t).$$

It follows that the Caputo derivative is not the mathematical extension of a typical derivative. More illustrations are given in [9].

2.5 Problems

1. [Fractional Stokes and Rayleigh Equations] The classical Stokes problem deals with the unsteady boundary layer flows induced in a semi-infinite viscous fluid bounded by an infinite horizontal disk, kept at the $x-y$ plane ($z=0$), due to non-torsional oscillations of the disk in its own plane with a given frequency, ω. When $\omega = 0$, the Stokes problem reduces to the classical Rayleigh problem where the unsteady boundary layer flow is generated in the fluid from rest by moving the disk impulsively in its own plane with constant velocity U. Solve the unsteady fractional boundary layer equation for the fluid velocity $u(z,t)$, given by,

$$\begin{aligned}
&\frac{\partial^\alpha u}{\partial t^\alpha} = \nu \frac{\partial^2 u}{\partial t^2}, \quad 0 < z < \infty, \quad t > 0, \\
&u(0,t) = Uf(t), \quad u(z,t) \to 0 \text{ as } z \to \infty, \quad t > 0, \\
&u(z,0) = 0 \text{ for all } z > 0,
\end{aligned}$$

where ν is the kinematic viscosity, U is a constant velocity and $f(t)$ is an arbitrary function of time t.

2. If $f(t) = \sqrt{t}$, find (a) Grünwald-Letnikov, and (b) L1 derivative approximation for fractional derivative $\alpha = \frac{1}{2}$.

References

1. Ochmann M, Makarov S (1993) Representation of the absorption of nonlinear waves by fractional derivatives. J Am Accost Soc 94(6):3392–3399
2. Miller KS, Ross B (1993) An introduction to the fractional calculus and fractional differential equations. Wiley-Blackwell
3. Podlubny I (1999) Fractional differential equations: an introduction to fractional derivatives, fractional differential equations, to methods of their solution and some of their applications. Academic Press, p 198
4. Li C, Zeng F (2015) Numerical methods for fractional calculus. CRC Press
5. Goychuk I, Pöschel T (2021) Fingerprints of viscoelastic subdiffusion in random environments: revisiting some experimental data and their interpretations. Phys Rev E 104(3):034125
6. Meerschaert MM, Scheffler HP, Tadjeran C (2006) Finite difference methods for two-dimensional fractional dispersion equation. J Comput Phys 211(1)
7. Diethelm K, Freed AD (2006) An efficient algorithm for the evaluation of convolution integrals. Comput Math Appl 51(1):51–72
8. Magin RL, Abdullah O, Baleanu D, Zhou XJ (2008) Anomalous diffusion expressed through fractional order differential operators in the Bloch-Torrey equation. J Magn Reson 190(2):255–270
9. Li CP, Zhao ZG (2011) Introduction to fractional integrability and differentiability. Eur Phys J-Special Topics 193:5–26

Chapter 3
Spatiotemporal Linear Stability Analyses

Abstract This chapter describes the temporal and spatiotemporal linear stability analyses of viscoelastic, sub-diffusive, plane Poiseuille flows obeying the Fractional Upper Convected Maxwell (FUCM) equation in the limit of low to moderate Reynolds number (Re) and low Weissenberg number (We). In particular, we demonstrate how the exponent in the sub-diffusive power-law timescale (t^{α}, with $0 < \alpha \leq 1$) at the microscale is related to the fractional order of the time derivative, α, of the corresponding nonlinear stress constitutive equation in the continuum. A description of the various kinds of temporal and spatiotemporal instability occurring in viscoelastic flows, namely the absolute instabilities, the convective instabilities and the evanescent modes, is provided. Results are limited to two exponents for the fractional order of time-derivative: monomer diffusion in Rouse chain melts, $\alpha = 1/2$, and in Zimm chain solutions, $\alpha = 2/3$. The temporal stability analysis indicates that with decreasing order of the fractional derivative: (a) the most unstable mode decreases, (b) the peak of the most unstable mode shifts to lower values of Re, and (c) the peak of the most unstable mode, for the Rouse mode, precipitates towards the limit $Re \to 0$. The Briggs idea of analytic continuation is deployed to demarcate the boundaries of the various types of instabilities. The spatiotemporal phase diagram indicates an abnormal region of temporal stability at high fluid inertia, revealing the presence of a non-homogeneous environment with hindered flow, thus highlighting the potential of the model to effectively capture certain experimentally observed flow-instability transition in sub-diffusive flows.

3.1 Introduction

Understanding the hydrodynamic stability and transition of viscoelastic melts is paramount to both the fundamental theory of viscoelastic liquids and their industrial applications. The study of spatiotemporal stability is concerned with how minor perturbations to the flow behave and whether they grow or decay through space and time. The non-dimensional numbers that govern the stability of fluids are the Reynolds number and Weissenberg number. The ratio of inertial and viscous forces, known as the Reynolds number, aids in the prediction of flow behaviour. When viscous

H. Wilson et al., *Viscoelastic Subdiffusive Flows*, Fluid Mechanics and Its Applications 138, https://doi.org/10.1007/978-981-96-0135-6_3

forces prevail at low Re, the flow is smooth, referred to as laminar flow, and when inertial forces dominate at high Re, the flow tends to create chaotic eddies, vortices, and other instabilities. The Reynolds number is a key metric in fluid mechanics that is used to anticipate the onset of turbulence in a variety of fluid systems, such as pipelines, channels, and boundary layers. In the study of non-Newtonian fluids, the Weissenberg number is crucial. When We is small, the elastic forces are weak in comparison to the viscous forces, and the fluid behaves like a Newtonian fluid. However, when We increases, so do the elastic forces, resulting in complicated flow behaviour such as flow instabilities, flow patterns, and vortex formation [1].

Researchers have utilized linear stability analysis to explore the start of instability in viscoelastic flows with various geometries, such as straight channels, curved channels, and circular flows. The emergence of complex spatiotemporal patterns in viscoelastic flows, such as oscillations, travelling waves, and chaos, have been studied using nonlinear stability analysis and numerical simulations [2]. Understanding the stability properties of viscoelastic flows can aid in the design of industrial processes involving complex fluid transport, such as polymer manufacturing and oil recovery [3]. Research has shown that mathematical models and numerical simulations are precise methods to examine the spatiotemporal stability of viscoelastic flows [4].

In this chapter, we have limited our focus on the linear, temporal and spatiotemporal analyses of viscoelastic, planar, Poiseulle flows. We demonstrate how the dynamics of the sub-diffusive fluids at the microscale (often represented via a generalized Langevin equation (GLE) at the molecular level, using a dissipative memory kernel) is 'upscaled' to a fractional viscoelastic stress constitutive equation at the continuum level [5]. Using fractional derivatives in a constitutive law is a coarse-grained approach to capture the relaxation spectrum of the polymeric liquid at finer scales. This chapter is organized as follows. Section 3.2 outlines the different types of local as well as global instabilities encountered in a flow, including temporal instability (Sect. 3.2.1), absolute and convective instabilities (Sect. 3.2.2) and evanescent modes (Sect. 3.2.3) and demonstrates the method to evaluate these instability modes, namely the Briggs method (Sect. 3.2.4.3). Section 3.3 describes the linear stability analyses of plane Poiseuille flows, including the mathematical model (Sect. 3.3.1), the normal mode expansion (Sect. 3.3.2), the numerical method (Sect. 3.3.3), the results (Sects. 3.3.5 and 3.3.6), followed with a brief discussion on the implication of these results in Sect. 3.4.

3.2 Instability Analyses: Local Versus Global

In many flows of interest, the mean-velocity profile is nonuniform in the streamwise direction, and in order to distinguish between the local instability and the global instability, it is first essential to assume that streamwise variations of the mean flow are slow over a typical instability wavelength (the so-called parallel flow approximation). The terms "local" and "global" then refer to the instability of the local velocity profile and of the entire flow field, respectively. In this chapter, we will further characterize the local impulse response of the system within the parallel-flow approximation at each streamwise station.

As a first step in comprehension, it is appropriate to briefly recall the classical hydrodynamic instability description of open flows (flows with domain boundaries at infinity such as shear flows, homogeneous mixing layers, flat-plate wakes, uniform-density jets [6]). Since several spatially developing shear flows are known to be extremely sensitive to external noise, many controlled experiments have been conducted to determine their response to different excitation frequencies. As a result, it has been customary to represent the downstream development of vortical structures as a collection of spatially growing instability waves of various frequencies (refer to the work of Ho and Huerre [2] for a review of such analyses applied to mixing layers). In other words, experimental observations have been compared with the results of local spatial stability calculations (with a given real frequency and unknown complex wave number) performed on the measured time-averaged mean velocity profile at each streamwise station [7]. Such an approach has been reasonably successful in describing the evolution of vortices in forced experiments. For examples of this type of analysis, the reader is referred to the references [7, 8]. A large part of the hydrodynamic stability literature, however, has been devoted to the temporal theory (with a given real wave number and unknown complex frequency), where it is implicitly assumed that the flow develops from some given initial state. The following question may then be asked: What is the fundamental reason for adopting a spatial theory point of view in many open flows? It is argued here that the notions of local "absolute/convective" instability provide a rigorous justification for selecting spatial theory in specific open flows. However, the application of these concepts to other open flows (bluff body wakes, heated jets) leads to the conclusion that local spatial theory in the strict sense (real frequency) is not appropriate [9]. Wave number and frequency both need to be considered as complex numbers, and a global temporal instability may arise whereby the entire nonparallel mean flow admits self-sustained global modes with well-defined complex frequencies [4].

The impact of such concepts on the study of fluid-mechanical instabilities appears to be of relatively recent origin. For example, a spatiotemporal description of Tollmien-Schlichting wave packets in boundary layers was developed early on by Gaster et al. [8], whereby a general formal methodology was proposed without explicitly introducing a definite distinction between the absolute or convective nature of the instability mechanism. The technique advocated by Briggs [10] has been repeatedly implemented to analyze the receptivity of compressible shear flows due to acoustic forcing. In geophysical fluid dynamics, Thacker [11] and Merkine [12] have determined the transition from absolute to convective instability in a two-layer model of baroclinic instability. The distinction between absolute and convective instabilities, especially in spatially evolving shear flows, appears to have first been brought out in a general context in the works of Twiss [13] and Landau [14]. Similarly, plasma physicists have made extensive and seminal contributions to the theoretical foundations underlying the notions of hydrodynamic instabilities, and they have applied them to the study of numerous plasma instabilities. For systematic developments of these ideas, the reader is referred to the work of Sturrock [15], Briggs [10], and Lifshitz [16]. The complete and lucid review of Bers [17] is particularly recommended for an up-to-date account of theoretical efforts in the description of spatiotemporal plasma instabilities. The topic of absolute/convective instabilities has found its way into the mainstream of the plasma-physics literature, and presentations of the main ideas have appeared in the book by Clemmow et al. [18].

Local/global and absolute/convective instability concepts provide the necessary theoretical framework to classify the different types of open and unsteady flows according to the qualitative nature of their dynamic behaviour [4]. The spatial evolution of these flows is largely determined by the character (amplitude, frequency content, etc.) of the excitation. For instance, shear flows that are locally convectively unstable everywhere (such as mixing layers) essentially display extrinsic dynamics. Such flows are 'noise amplifiers'. In contrast, shear flows with a pocket of absolute instability of sufficiently large size (for example, bluff-body wakes, hot or low-density jets) display intrinsic dynamics of the same nature as in the corresponding closed-flow systems. Fluid particles are still advected downstream, but temporally growing global modes may be present. These flows behave as oscillators in the sense that the evolution of vortices does not rely on the spatial amplification of external perturbations but rather on the growth of initial disturbances in time. Furthermore, the distribution of global modes is synchronized in the streamwise direction. This class of open flows is particularly well suited to a nonlinear dynamical systems approach of the kind that has been implemented in closed systems. The onset of deterministic chaos, if it exists, is likely to be well-defined in these systems: one expects a well-ordered sequence of bifurcations from a limit cycle (a global phenomenon) to a low-dimensional strange attractor. The sensitivity of convectively unstable flows to external noise, on the other hand, makes it much more difficult to discriminate between low dimensional chaos arising from the flow dynamics and spatially amplified random noise. Finally, there exists a third class of marginally globally stable modes, the so-called 'evanescent modes', where the local velocity profiles are, strictly speaking, locally convectively unstable in the entire field, but absolute instability is incipient at some streamwise station. In such situations, global modes are often weakly damped in time, and they can be preferentially destabilized by applying external forcing in the vicinity of the global mode frequency. An in-depth discussion on Temporal Instability, Absolute/Convective Instability and the concept of Evanescent modes are discussed in the next three subsections. In order to numerically evaluate these instabilities, the key concepts of branch/pinch points, the Cusp-Map diagram, as well as the Briggs' Contour Integral Method, elaborated through an example, are delineated in the subsequent four subsections.

3.2.1 Temporal Instability

The classical linear stability theory of parallel flows is concerned with the development in space and time of infinitesimal perturbations around a given basic (mean) flow $U(y; Re)$. In the sequel, x, y, and t denote the streamwise direction, cross-stream direction, and time, respectively, and $U(y; Re)$ is the sole component of the basic flow in the x-direction. The basic state is parallel, i.e. it is assumed to be independent of x. Fluctuations are typically decomposed into plane wave instability of the form: $\phi(y; \alpha)exp[i(\alpha x - \omega t)]$ with complex wave number α and complex frequency ω. The cross-stream distribution $\phi(y; \alpha)$ is then shown in most cases to satisfy an ordinary differential equation of the Orr-Sommerfeld type [4]. Enforcement of appropriate boundary conditions leads to an eigenvalue problem whereby the eigenfunctions $\phi(y; \alpha)$ exist only if α and ω are constrained to satisfy a dispersion relation of the form,

$$D[\alpha, \omega, \boldsymbol{M}] = 0, \tag{3.1}$$

where $\boldsymbol{M}$ is the vector of material and fluid parameters. For simple basic flows (or flows with a known functional form of the mean flow, $U(y; Re)$), this relation may be found explicitly. For more realistic velocity profiles, the dispersion relation is obtained by the numerical integration of the Orr-Sommerfeld equation. In our subsequent discussion, we denote the real (imaginary) component of a variable with subscript, r (i), respectively. *Temporal modes*, $\omega(\alpha, Re)$, refer to cases where the complex frequency ω is determined as a function of the real wave number, α, and the temporal growth rate, ω_i^{Temp}, is the largest positive imaginary component of the dispersion relation (3.1). Conversely, the spatial branches $\alpha(\omega; Re)$ are obtained by solving for the complex wave numbers, α, when ω is given real. In this section, we deliberately ignore variations of the plane wave instability in the cross-stream direction, y, and only consider the spatiotemporal evolution of instability waves in the (x, t)-plane. This projection greatly simplifies the presentation of the fundamental concepts without losing any of the essential characteristics of the instability. Thus, one may associate an integro-differential operator $D[-i(\partial/\partial x), i(\partial/\partial t); \boldsymbol{M}]$ in physical space (x, t) corresponding to the dispersion relation (3.1) in spectral space $(\boldsymbol{M}, \omega)$, such that fluctuations $u(x, t)$ satisfy

$$D\left[-i\frac{\partial}{\partial x}, i\frac{\partial}{\partial t}; \boldsymbol{M}\right] u(x, t) = 0. \tag{3.2}$$

To solve the corresponding receptivity problem for the above equation, one introduces an impulse response given by the Green's function, $G(x, t)$, as follows,

$$D\left[-i\frac{\partial}{\partial x}, i\frac{\partial}{\partial t}; R\right] G(x, t) = \delta(x)\delta(t), \tag{3.3}$$

with δ denoting the Dirac delta function source. The basic flow is then said to be linearly stable if

$$\lim_{t\to\infty} G(x, t) = 0 \text{ along all rays where } x/t = \text{ constant }, \tag{3.4}$$

and it is linearly unstable if

$$\lim_{t\to\infty} G(x,t) = \infty \quad \text{at least along one ray where } x/t = \text{ constant.} \tag{3.5}$$

3.2.2 *Spatiotemporal Instability: Absolute Versus Convective Modes*

The spatiotemporal analysis is typically relevant when one introduces an impulse excitation locally in a flow and observes how the disturbance evolves. Among linearly unstable flows (3.5), one must further distinguish between two types of impulse response: the basic flow is referred to as convectively unstable if,

$$\lim_{t\to\infty} G(x,t) = 0 \text{ along the ray } x/t = 0, \tag{3.6}$$

and it is absolutely unstable if,

$$\lim_{t\to\infty} G(x,t) = \infty \text{ along the ray } x/t = 0, \tag{3.7}$$

as shown in Fig. 3.1. Equations (3.6) and (3.7) imply that if the localized disturbances are swept away from the source, the velocity is said to be locally convectively unstable. In contrast, if the localized disturbances spread upstream and downstream and contaminate the entire parallel flow, the velocity profile is said to be absolutely unstable in the local sense.

Physically, convectively unstable flows give rise to wave packets that move away from the source and ultimately leave the medium in its undisturbed state. Absolutely unstable flows, by contrast, are gradually contaminated everywhere by a point-source input. In order to distinguish between convective and absolute instabilities, it is further

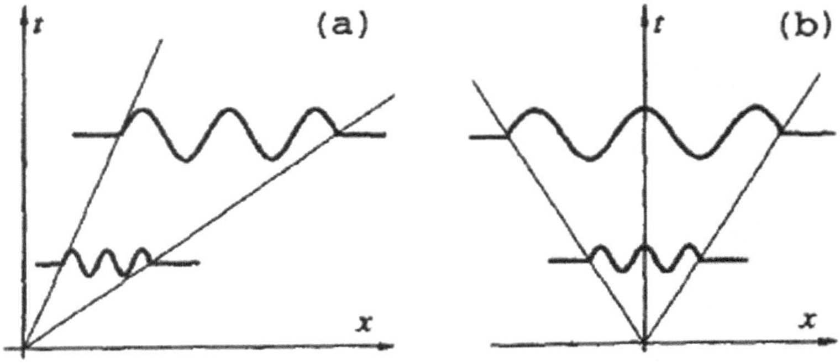

Fig. 3.1 **a** Convective, **b** Absolute instability. (*Source* [4])

necessary to examine the long-time behaviour of the wave number α_0 observed along the ray $x/t = 0$ at a fixed spatial location. This complex wavenumber, α_0, has, by definition, a zero group velocity, or,

$$\frac{\partial \omega}{\partial \alpha}(\alpha_0) = 0, \tag{3.8}$$

where the corresponding frequency, $\omega_0 = \omega(\alpha_0)$, is commonly called the absolute frequency. The absolute growth rate, ω_i^{Cusp}, (or the growth rate at the cusp point, see Sect. 3.2.4 for details) is then denoted by, ω_i^{Cusp} = Imag($\omega(\alpha_0)$). In other words, the absolute growth rate, ω_i^{Cusp}, characterizes the temporal evolution of the wave number α_0 observed at a fixed station in the limit $t \to \infty$. The sign of ω_i^{Cusp} determines the absolute/convective nature of the flow, leading to the following criterion:

$$\begin{aligned} \omega_i^{\text{Cusp}} &> 0 \quad \text{absolutely unstable flow,} \\ \omega_i^{\text{Cusp}} &< 0 \quad \text{convectively unstable flow.} \end{aligned} \tag{3.9}$$

3.2.3 *Evanescent Modes*

A flow field governed by the Navier-Stokes equations can sustain, in general, three different types of modal wave fields, namely, acoustic, vorticity and entropy waves [19]. In the linear approximation, these modal fields are uncoupled except at boundaries or possibly at direct resonances. One speaks of a 'direct resonance of evanescent modes' if (in a physical system that allows more than one wave mode) two or more of these modes coalesce. In the present investigation, we shall concentrate on the coalescence of modes of the same wave type and, in particular, on the coalescence of vorticity modes. If none or only very weak instabilities are present, then direct resonance can be a rather powerful selection mechanism leading to an algebraic growth for short times or at short distances, and the corresponding potentially large amplitudes may initiate nonlinearity, leading to a finite time 'blowup' long before the exponentially growing mode does.

The concept of direct resonance through evanescent modes is demonstrated in Fig. 3.2 by plotting the functions $f(x) = x\exp(-\alpha_i x)$ versus $f(x) = \exp(\alpha_i x)$ for various α_i =Imag(α). For large α_i high amplitude $f(x)$ are quickly reached by the exponentially growing mode, $\exp(\alpha_i x)$. However, if α_i is small, this growth is very slow, and therefore, a direct resonance generating function, $x\exp(-\alpha_i x)$ can produce large amplitudes much faster. Although these modes ultimately decay according to linear theory, the locally high amplitudes may initiate the nonlinear state. In order to determine if this is possible for a given flow, we have to find out (i) under what conditions direct resonances exist and (ii) how large the corresponding damping rates, α_i, are.

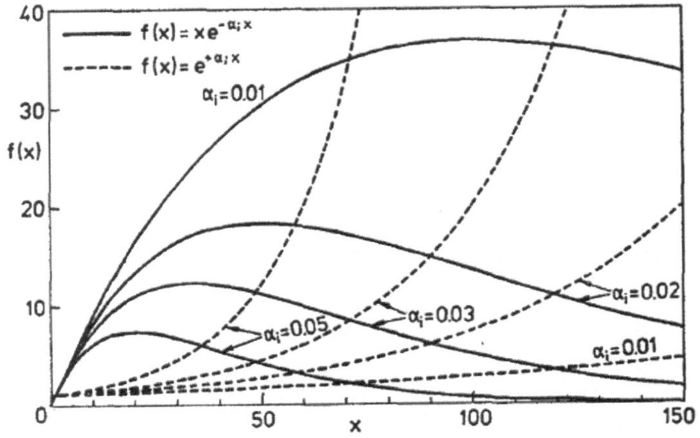

Fig. 3.2 Example of an algebraic function elucidating evanescent mode. (*Source* [20])

The concept of direct resonance has significant consequences in the generation of waves by wind [21] and in the stability of locally parallel shear flows (refer to these references: [22–26] for details on a weakly nonlinear theory of waves at direct resonance). In the above-cited articles concerning the stability of parallel shear flows, the resonant forcing of the vertical vorticity (and hence the horizontal velocity) components by the vertical velocity has been explored with the goal of elucidating the laminar-turbulent transition process. This corresponds to a direct resonance between an Orr-Sommerfeld and a vertical vorticity or so-called 'Squire' mode. However, the main interest of this chapter lies in finding out whether the flow can support a direct resonance between two Orr-Sommerfeld modes alone and whether such a resonance mechanism can be of physical importance. The character of a direct resonance depends on the type of singularity in the dispersion relation at the point of mode coalescence. It is usually determined by studying the response to a point source in space and time. If the coalescing modes originate from waves double pole, the singularity is of the 'pinch'-type. For example, in a linear neutrally stable system, the 'cut-off' phenomenon separating propagation from an evanescent acoustic wave in a hard-walled duct belongs to this category [27]. If the two coalescing modes originate from waves propagating in the same direction as in the case of a convectively unstable flow, the corresponding singularity is of the 'double pole' type. Such a disturbance will ultimately decay according to linear theory. However, the short-term algebraic growth associated with such a double pole may be decisive and carry the system to a nonlinear state. Whether this is physically possible is determined by the damping rate associated with the mode coalescence. If the damping rate is large, such resonances are of importance in certain linear optimization problems (an example can be seen in a work by Cremer [28] for ducts with acoustically absorbing walls). The corresponding damping rates are very small, and the algebraic growth may dominate the exponential decay locally. This is the case addressed by Benny et al. [29] as well as in the present investigation of plane Poiseuille flows and boundary-layer type mean flows, which are typical examples of convectively unstable flows.

3.2.4 *Evaluation of Instability Modes*

It is the goal of this section to introduce a method for detecting absolute instabilities (and consequently convective instabilities and evanescent modes) that solely relies on a mapping from the α-plane to the ω-plane. This technique is originally attributed to Kupfer et al. [30] and referred to as the 'cusp map' in the complex-frequency plane. We start with the solution to the stability problem in Fourier-Laplace space. If we use Fourier- and Laplace-transform for Eq. (3.3) and formally revert back to the physical space, we have,

$$G(x,t) = \frac{1}{4\pi^2}\int_L\int_F \frac{\exp[i(\alpha x - \omega t)]}{D(\alpha,\omega,\boldsymbol{M})} d\alpha d\omega, \tag{3.10}$$

where L and F denote the inversion contour in the Laplace-ω-plane and the Fourier α-plane, respectively. Although the Fourier-Laplace integral (3.10) could be determined for all x and t, the complexity of the dispersion relation suggests a time-asymptotic solution of the integral. Moreover, the definition of convective and absolute instability as a limiting process for large times also warrants an asymptotic approach in evaluating the Fourier-Laplace integral (3.10). The time-asymptotic evaluation of the integral expression for Green's function will involve the deformation of the integration contours in the complex α and ω planes. By deforming the integration path through a saddle point in the plane, we can use the method of steepest descent to evaluate the time-asymptotic behaviour. If this evaluation results in a divergent integral, the flow under investigation is considered absolutely unstable according to the definition. If the asymptotic limit results in a convergent integral (in fact, in a zero integral), the flow is convectively unstable [31]. Careful evaluation of the time-asymptotic behaviour of the Fourier-Laplace integral (3.10) is accomplished using a method introduced by Briggs [10] (described in Sect. 3.2.4.3).

3.2.4.1 Branch Points and Pinch Points

A pinch point, α^{pinch}, in the complex α-plane (which corresponds to a branch point in the complex ω-plane, $\omega^{Branch} = \omega^{\text{Cusp}}$) is a special kind of saddle point satisfying the relation,

$$D\left(\alpha^{pinch},\omega^{\text{Cusp}}\right) = 0 \quad \frac{\partial D}{\partial \alpha}\left(\alpha^{pinch},\omega^{\text{Cusp}}\right) = 0 \quad \frac{\partial^2 D}{\partial \alpha^2}\left(\alpha^{pinch},\omega^{\text{Cusp}}\right) \neq 0. \tag{3.11}$$

The second condition in equation (3.11) is the saddle point condition in the complex plane, which equates to zero group velocity, where the group velocity V_g is such that,

$$V_g = \frac{\partial \omega}{\partial \alpha} = \frac{\partial D}{\partial \alpha} / \frac{\partial D}{\partial \omega}. \tag{3.12}$$

The third condition in Eq. (3.11) implies that the branch point in the complex ω-plane has only $n - 1$ image points corresponding to n branches, $\alpha_k(\omega)$, such that two sheets in the ω-plane connect at a specific point, to form a 'cusp'. In order to understand the cusp-point formation, consider the Taylor series expansion of the dispersion relation (3.1) in the neighbourhood of the singularity (α^{pinch}, ω^{Cusp}), which leads to,

$$0 = \left.\frac{\partial D}{\partial \omega}\right|_0 \left(\omega - \omega^{\text{Cusp}}\right) + \frac{1}{2} \left.\frac{\partial^2 D}{\partial \alpha^2}\right|_0 \left(\alpha - \alpha^{pinch}\right)^2 + \mathcal{O}\left(\left(\omega - \omega^{\text{Cusp}}\right)^2, \left(\alpha - \alpha^{pinch}\right)^3\right). \tag{3.13}$$

This expression gives a relation between the neighbourhood of ω^{Cusp} in the ω-plane and the neighbourhood of α^{pinch} in the α-plane. The quadratic nature of the local map between the α and ω planes results in a square root singularity for the local map between the ω and α planes. Generalizing this concept in the n-dimensional plane, the local map between the α and ω plane shows an $(n - 1)$th root singularity. This local singularity forms the basis of the cusp map procedure, described next.

3.2.4.2 The Cusp Map

The asymptotic space-time evolution of a linear instability is governed by the location of pinch points in the complex α-plane that is found by an analytic continuation of the Laplace inversion contour (3.10) into the lower half of the ω-plane. This technique requires solving the dispersion relation for the spatial wavenumber α as a function of the frequency ω. In general, dispersion relations are higher order polynomials (or even transcendental functions) in the wave number α, and thus, the mapping of the F-contour into the ω-plane is multivalued. Let us denote the image of F in the ω-plane by $\omega(F)$. The reverse mapping of $\omega(F)$ back into the α-plane will result in n branches, $\alpha_k(\omega(F))$, with n as the highest power of α in the dispersion relation (assumed polynomial). Nevertheless, the original contour F is recovered by one of the n branches. To make the mapping of the F-contour into the ω-plane single-valued, we introduce n Riemann sheets in the complex ω-plane and associate the n contours in the multi-sheeted ω-plane with the n branches, $\alpha_k(\omega(F))$, in the α-plane. The map between the ω and α-plane is governed by the dispersion relation $D(\alpha, \omega) = 0$. For a pinch point in the complex α-plane, we have the additional constraints, which are outlined in Eq. (3.11).

We will now concentrate on these two Riemannian sheets that connect at a specific branch point leading to the formation of the cusp point (or the branch point satisfying all the constraints in Eq. (3.11)). Let us introduce a vertical ray that connects the contour L in each of the two sheets to the point ω_0 (refer to Fig. 3.3). We will call

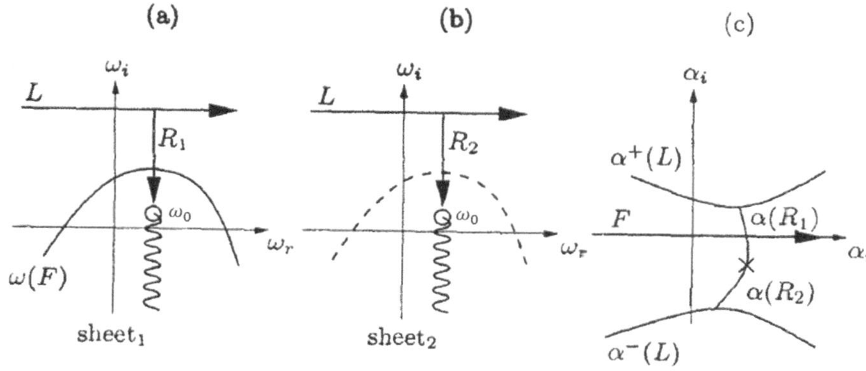

Fig. 3.3 Sketch illustrating the cusp map method. (*Source* [30])

these rays R_1 and R_2. They will help us determine whether the point ω_0 in the ω-plane corresponds to a pinch point α^{pinch} in the α-plane. This happens if and only if the images of the rays R_1 and R_2 in the complex α-plane, i.e., $\alpha(R_1)$ and $\alpha(R_2)$ in Fig. 3.3, originate on two different sides of F. If the images $\alpha(R_1)$ and $\alpha(R_2)$ originate on different sides of F but connect at one single point α^{pinch}, which is not on F, we must conclude that at least one of these images crosses F. If more than one crossing occurs, the total number of crossings by both images must be odd.

This criterion forms the basis of the cusp map procedure: by counting the number of times that each vertical ray, R_1 and R_2, intersects the contour $\omega_k(F)$ (which is the solution to the dispersion relation (3.1) corresponding to the n-branches, $\alpha_k(F)$, in the α-plane, refer Fig. 3.4), we can determine whether the branch point ω_0 corresponds to a pinch point in the α-plane. In a similar way, the stability characteristics of any dispersion relation may be determined. The procedure requires mapping a section of the α-plane into a specified region located beneath the ω-contour, $\omega(F)$. A set of vertical rays along the range of unstable wave numbers is mapped into the ω-plane. These rays, originally parallel in the α plane, may have images that intersect, thus indicating a branch point. The singularity is identified by the angle-doubling property of its local map. This is the case in Fig. 3.4, which shows the branch point nested at the edge of a cusp-like trajectory. If a branch cut is taken downward from the singularity, one obtains a mapping consistent with the multi-sheeted structure implied by the contour $\omega(F)$. In this case, the branch point is covered only once by $\omega(F)$ and thus corresponds to a pinch point. In many problems, it is simpler to replace the vertical rays shown in Fig. 3.4 with a set of horizontal contours that represent deformations of the Fourier integral path. In the ω-plane, these contours will progress downward from the map $\omega(F)$ and form a cusp as they approach the singularity; see Fig. 3.4. Once again, the branch point is located by the angle-doubling (or tripling, etc.) property of its local map. This procedure can be systematically generalized for dispersion relations with multiple unstable branches.

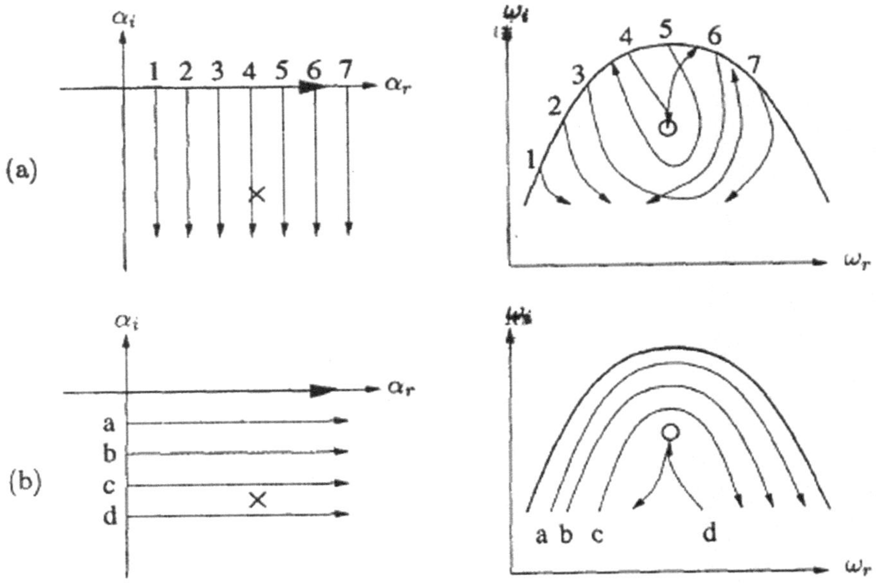

Fig. 3.4 Outline of the numerical procedure for detecting branch points in the ω-plane used in the cusp map method. (*Source* [30])

3.2.4.3 Briggs' Method

In the Briggs' method for instability classification [10], the wave number integral in equation (3.10) is chosen first,

$$\tilde{G}(x,\omega) = \frac{1}{2\pi}\int_F \frac{\exp[i\alpha x]}{D(\alpha,\omega,\boldsymbol{M})}d\alpha, \tag{3.14}$$

which is followed by the frequency component inversion,

$$G(x,t) = \frac{1}{2\pi}\int_L \tilde{G}(x,\omega)\exp[-i\omega t]d\omega. \tag{3.15}$$

The method of analytic continuation is used to deflect the original L-contour in an attempt to lower it below the real ω-axis (refer to the left column in Fig. 3.5). If this is achieved, the exponent in the ω-inversion integral (3.15) forces the integrand to vanish as $t \to \infty$; if not, the time-asymptotic discrete response is governed by the highest discrete singularity in the ω-plane. Lowering the ω-contour could fail if a singularity above the real ω-axis is encountered. Because of the interconnection through the dispersion relation, this singularity in the ω-plane will have an associated singularity in the α-plane (refer to the right column in Fig. 3.5).

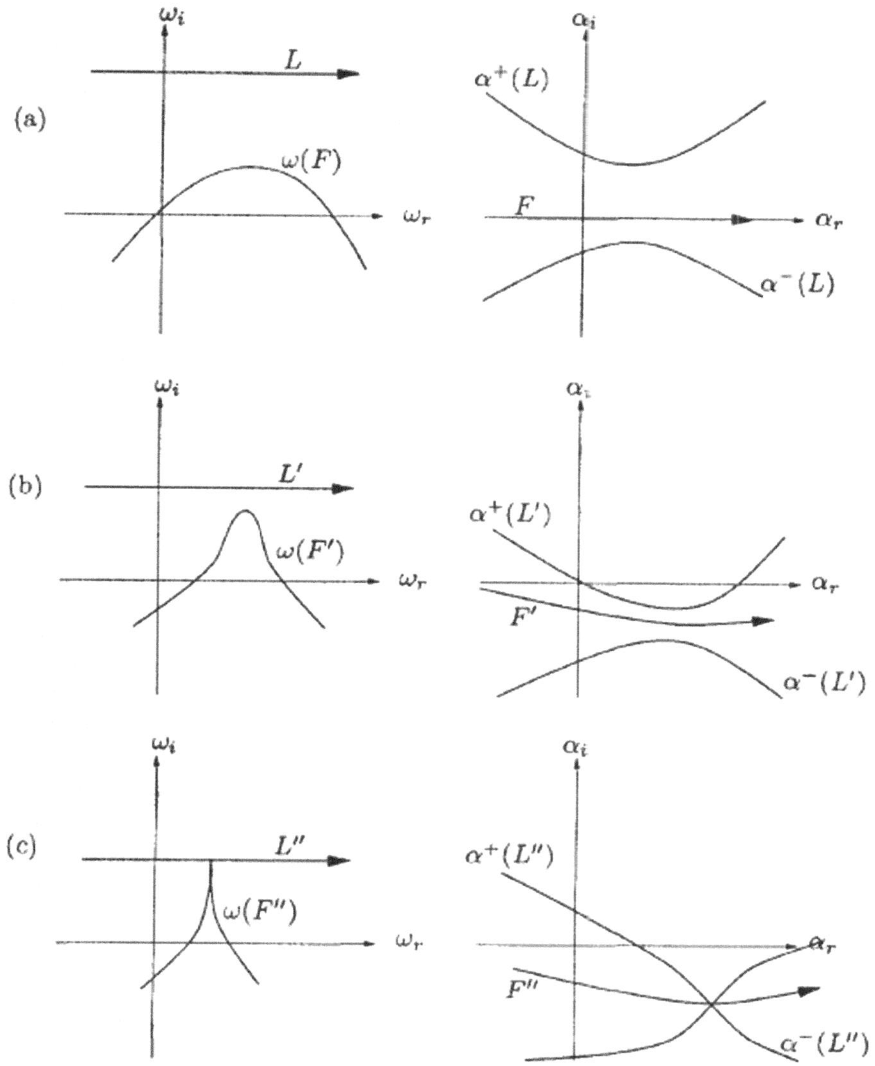

Fig. 3.5 Sketch of contour deformation procedure for Briggs' method. Left: complex ω-plane, right: complex α-plane. (*Source* [4])

We choose the real α-axis as the inversion contour for the spatial part. This integration path maps through the dispersion relation (3.1) to a curve in the ω-plane denoted by $\omega(F)$, as shown in Fig. 3.5. The temporal inversion contour (labelled 'L' in Fig. 3.5) has to lie above this curve in order to satisfy causality: the integration path for $t < 0$, which involves closure in the upper half-plane, cannot encircle any singularities, or, in other words, the integrand of the ω integral (3.15) has to be analytic in the ω-half-space: $\omega_i > \max \text{Imag}(\omega(F))$. The temporal contour L can also be mapped back into the α-plane, resulting in the branches labelled $\alpha^+(L)$ and $\alpha^-(L)$. The spatial branch above (below) the real axis is associated with the dynamics downstream (upstream) of the source, respectively. As we deform the ω-contour, so as to lower it below the real axis, its image in the α-plane will also deform. At some point, the original α-contour will be squeezed between the two branches, which will necessitate the deformation of the original ω-contour. By constantly adjusting the inversion contours and their maps into the associated plane, we may arrive at a situation where the spatial inversion contour is pinched between the two spatial branches, therefore prohibiting any further deformation without crossing singularities. If this occurs, the temporal inversion contour cannot be lowered any further due to the formation of a cusp point in the ω-plane (see Fig. 3.5c).

An absolute instability is formed from the merging of two eigenmodes (or the $\omega(F)$ contours). As such, its occurrence may be viewed intuitively as a form of resonance between the modes; a resonance, which leads to growth, since the merged eigenmode has $\omega_i > 0$. Physically, the modal interaction associated with the merging takes place across the source of the perturbation. This is because the participating modes are from the upstream and the downstream sides of the source. Thus, the instability is localized with respect to the location of the source and envelopes an ever-increasing neighbourhood of the perturbation source with time. A shear flow with a sizable pocket of absolute instability may display the dynamic characteristics of a closed-flow (compact) system. These unstable flows behave like an oscillator. A convectively unstable mode, on the other hand, has been shown to be associated with a sinusoidal wave that grows in amplitude as it travels away from the source, from which it also derives its frequency. A convectively unstable mode is essentially a driven mode. It decays to zero with time when the driving source is turned off, whereas an absolutely unstable mode is more akin to a self-sustaining temporal mode and has been termed as such by Bers [17]. Convectively unstable modes are produced by α-roots, which end up in a different half of the α-plane from the one in which they originate when the L contour is deformed onto the real α-axis. The crossing of the α roots across the real axis distinguishes them from evanescent modes. A fuller account of the theory is given by Bers [17].

When $\omega_i^{\text{Cusp}} > 0$, a possible absolute instability is indicated. It is still necessary, however, to verify that the intersection arises from roots originating from opposite halves of the α-plane. For such an intersection to be possible, at least one of the α-roots must cross the α_r-axis of the α-plane at least once. An even number of crossing brings an α-root back into its original half of the α-plane, while an odd number brings it into the opposite half. The sum total of crossings must be an odd number for a genuine pinch point. The fulfilment of this requirement can be checked by drawing a straight ray from the suspected cusp point vertically upward (ω_r = constant) and observing the number of times this ray intersects the image of the α_r-axis ($\alpha_i = 0$) in the ω-plane (a double or multisheeted Riemann surface). This is because every crossing of the α_r-axis in the α-plane by the α-roots (as ω_i is varied with the lowering

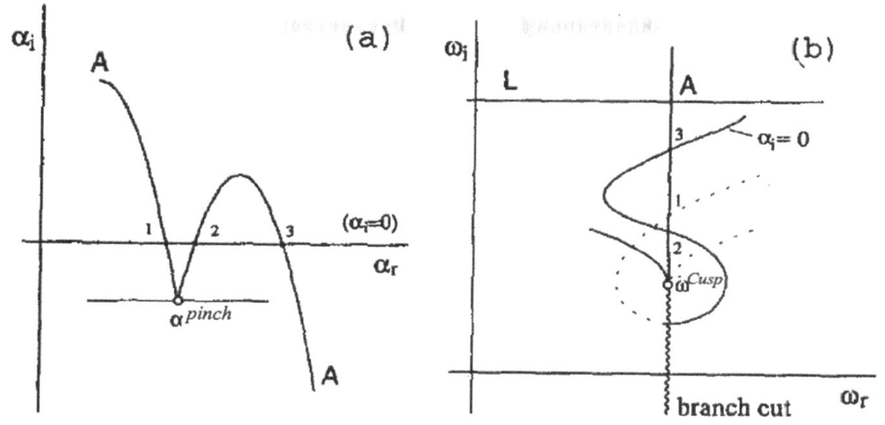

Fig. 3.6 An example in the schematic of an absolute instability created by the intersection of two α-roots with three crossing of the α_r-axis ($\alpha_i = 0$). The vertical ray from ω_0 cuts the $\alpha_i = 0$ contour in the three places in the double-sheeted ω-plane. (*Source* [32])

of the L contour) is reflected by a corresponding crossing of the image of the α_r-axis ($\alpha_i = 0$) in the ω-plane by the vertical ray (which represents variation in ω_i) in one of the Riemann sheets of the co-plane. Figure 3.6 presents a schematic example of the application of this criterion to a situation in which the intersecting pair of α-roots crosses the α_r-axis a total of three times. As shown in this figure, the upward vertical ray emanating from the cusp point cuts the image of the α_r-axis ($\alpha_i = 0$) three times on the two Riemann sheets. When the Riemann sheets of the branch point, ω^{Cusp}, correspond to spatial branches located on the same side of α-plane for high enough L, no pinching of F can occur as L is lowered. The corresponding branch point in the ω-plane is not related to the absolute growth rate. Thus, care must be exercised to locate branch-point singularities pertaining to spatial branches originating from distinct halves of the α-plane. Thus, the vertical ray criteria can be used to distinguish between genuine absolute and convective instabilities from evanescent modes [32].

We summarize the Briggs' method as follows: The necessary (but not sufficient) condition for the presence of absolute instability is the vanishing characteristic of the group velocity, V_g, at the saddle point in the α-plane or the branch point in the ω-plane ($V_g = \frac{\partial \omega}{\partial \alpha} = (\frac{\partial D}{\partial \alpha})/(\frac{\partial D}{\partial \omega}) = 0$ such that $\omega = D(\alpha)$). But the group velocity is zero at every saddle point, in particular where the two α-branches meet, independent of whether the branches originate from the same half of the α-plane (i. e., when evanescent modes are detected) or not. To overcome this inadequacy, Briggs [10] devised the idea of analytic continuation in which the Laplace contour L, in equation (3.10), is deformed towards the ω_r axis of the complex ω-plane, with the simultaneous adjustment of the Fourier contour F in the α-plane to maintain the separation of the α-branches; those which originate from the top half (the upstream modes with $\alpha_i > 0$) from those which originate from the bottom half of the α-plane (or the downstream modes). The deformation of the F contour (while preserving causality) is inhibited, however, when the paths of the two α-branches originating from the opposite halves of the α-plane intersect each other, leading to the appearance of saddle points, which are the pinch point, α^{pinch}. The concurrent branch point appearance in the ω-plane is the cusp point, ω^{Cusp} (i. e., $D(\alpha^{\text{pinch}}, \omega^{\text{Cusp}}) = \frac{\partial D(\alpha^{\text{pinch}}, \omega^{\text{Cusp}})}{\partial \alpha} = 0$ but $\frac{\partial^2 D(\alpha^{\text{pinch}}, \omega^{\text{Cusp}})}{\partial \alpha^2} \neq 0$). Kupfer [30] employed a local mapping procedure to conceptualize the stability characteristics of this branch point. Near a 'reasonably close' neighbourhood of the pinch point, a local Taylor expansion yields a dispersion relation that has a second-order algebraic form in the ω-plane (and which is a first-order saddle point in the α-plane), i. e., $(\omega - \omega^{\text{Cusp}}) \sim (\alpha - \alpha^{\text{pinch}})^2$. This period-doubling characteristic of the map causes the α_i-contours to 'rotate' around ω^{Cusp}, forming a cusp. In the ω-plane, we draw a ray parallel to the ω_i-axis from the cusp point such that it intersects the image of the F-contour (or $\alpha_i = 0$ curve) and count the number of intersections (consequently, count the number of times both α-branches cross the α_r-axis before forming a pinch point in the α-plane, as shown in Fig. 3.6). If the ray drawn from the cusp point intersects the image of the F contour in the ω-plane (or if either one or both the α-branches cross the α_r-axis) even number of times, then the flow dynamics correspond to an evanescent mode. Otherwise, in the case of odd intersections, the observed cusp point is genuine, leading to either an absolutely unstable system (in the upper half of the ω-plane) or a convectively unstable system (in the lower half of the ω-plane); provided the system is temporally unstable.

3.2.4.4 An Example

Further demonstration of the Briggs' method is exemplified through a simple dispersion relation [30]. The dispersion relation outlined in this example is a mathematical one. Its simplicity allows explicit solutions for the pinch point and branch point locations, shown below,

$$D(\omega, \alpha) = \omega - \left[\frac{1}{3}(\alpha - i)^3 + i - \alpha V \right], \tag{3.16}$$

where we have also introduced a parameter V. The pinch points and branch points can be found by solving $d\omega/d\alpha = 0$, which yields,

$$\alpha_{1,2} = i \pm \sqrt{V}, \tag{3.17}$$

for the pinch points and,

$$\omega_{1,2} = (1 - V)i \mp \frac{2}{3} V \sqrt{V}, \tag{3.18}$$

for the associated branch points in the ω-plane. We set V = 0.75 and proceed with Briggs' method as outlined earlier. We map lines of constant ω into the complex α-plane by solving the cubic dispersion relation (3.16).

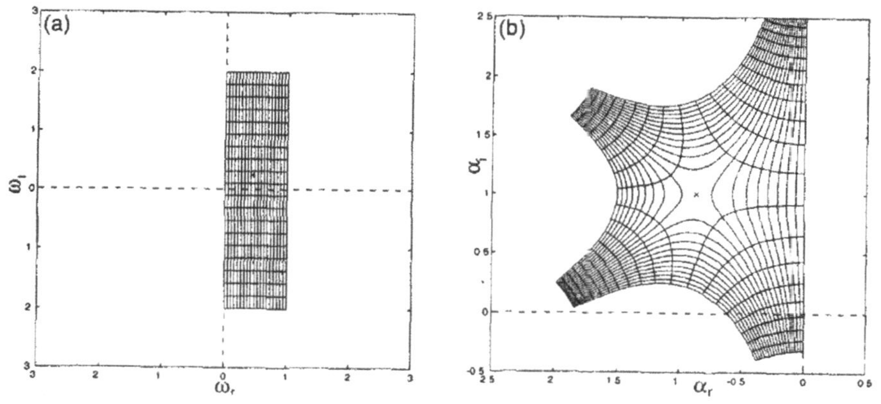

Fig. 3.7 Map of the complex ω-plane onto the complex α-plane under the dispersion relation (3.16). (*Source* [4])

Clearly, a saddle point forms in the complex α-plane that is marked by a symbol, 'x', and corresponds to the solution of equation (3.17) for V = 0.75 (refer Fig. 3.7). Its branches originate in different half-spaces of the α-plane. The corresponding branch point in the ω-plane is also marked. We see that the branch point is clearly above the real ω-axis, proving the occurrence of absolute instability. Admittedly, this example is fairly simplistic. It nevertheless demonstrates the usefulness of Briggs' method to detect absolute instabilities. More complicated dispersion relations can be contrived that challenge the readers' knowledge of analytic function theory. For a more detailed treatment, including numerical examples, the reader is referred to the work by Lingwood [31].

3.3 Linear Stability Analyses for Sub-diffusive Planar Poiseuille Flow

We now return to the problem of temporal and spatiotemporal linear stability analyses of viscoelastic, sub-diffusive, plane Poiseuille flow in the limit of low to moderate Reynolds number and low Weissenberg number. In the development of the mathematical model for the planar viscoelastic channel flow, we demonstrate how the exponent of the power-law scaling (t^{α}, with $0 < \alpha \leq 1$) in viscoelastic microscale models [33] is related to the fractional order of the time-derivative, α of the corresponding nonlinear stress constitutive equation in the continuum [5]. This significantly differs from the existing studies in the sense that we analyse the linear stability of viscoelastic, sub-diffusive channel flows through a combined temporal and spatiotemporal stability analysis (rather than only a temporal stability analysis of the classical (or integer order) viscoelastic channel flows [34]) and the aim is to address the following intriguing questions: What is the critical flow/polymer relaxation condition for the onset of instability? More crucially, what is the linear spatiotemporal, time asymptotic response of the flow at the critical value of the material parameters, leading to the topological transition of the advancing flow interface of the sub-diffusive channel flows? In particular, the temporal and the spatiotemporal stability of two specific cases of monomer diffusion in coarse-grained Rouse chain melts ($\alpha = 1/2$) [35], and in coarse-grained Zimm chain solution ($\alpha = 2/3$) [36] are discussed in detail.

3.3.1 Mathematical Model

We consider a viscoelastic fluid subject to shear deformation. Then, an infinitesimal elastic stress, τ_{xy} at time t arising from a small strain increment $d\gamma$ at an earlier time t' is given by,

$$d\tau_{xy} = G(t - t')\dot{\gamma}(t')dt', \tag{3.19}$$

where the relaxation modulus, $G(t)$, represents the influence of the dissipative processes of the surrounding concentrated fluid medium [37]. Assuming linearity, the Boltzmann superposition principle may be utilized to construct the elastic stress at time t by summing up all of the infinitesimal contributions over the entire flow history, which is extended into the infinite past [38],

$$\tau_{xy} = \int_{-\infty}^{t} G(t - t')\dot{\gamma}(t')dt'. \tag{3.20}$$

In their seminal work on passive micro-rheology, Mason and co-workers [33] have identified an approximate relation between the time-dependent memory kernel describing the viscous damping of the tracer particle at micro-scale (and which obeys the Generalized Langevin Equation, e. g., see equation (15) in [39]), $\zeta(t)$, and the stress relaxation modulus, $G(t)$, i. e.,

$$\zeta(t) = 6\pi a G(t), \tag{3.21}$$

where a is the radius of the tracer particle (assumed spherical). In the regime of linear viscoelasticity, one of the most commonly used three-parameter family of memory kernel is the generalized Rouse kernel for an equally weighted sum of negatively decaying exponential functions [40],

$$\zeta(t) = \frac{1}{N}\sum_{k=0}^{N-1} e^{-\left(\frac{k}{N}\right)^{\frac{1}{\alpha}}\left(\frac{t}{\lambda_0}\right)}, \tag{3.22}$$

for a number of kernels determining the length of the sub-diffusive phase, N, the shortest memory time scale, λ_0, and a sub-diffusive exponent, $\alpha \in (0\ 1]$. Since the polymeric liquids of our interest [41] show sub-diffusive behaviour on all length scales, we consider the case when $N \to \infty$ in the Prony series (3.22). For $t \gg \lambda_0$, this limiting behaviour leads to the relation,

$$\zeta(t) = \frac{\tilde{G}}{\Gamma(1-\alpha)} \left(\frac{t}{\lambda_0}\right)^{-\alpha}, \tag{3.23}$$

where $\Gamma(x)$ is the complete gamma function and $\tilde{G} = \Gamma(1+\alpha)\Gamma(1-\alpha)$, is a constant. In Eq. (3.23), we have used the fact that the Riemann sum on an infinite interval is given by,

$$\lim_{N\to\infty} \frac{1}{N} \sum_{k=0}^{N-1} f\left(\frac{k}{N}\right) = \int_0^{\infty} f(x)dx. \tag{3.24}$$

Using Eqs. (3.20, 3.21, 3.23), one arrives at,

$$\tau_{xy} = \frac{\mathcal{A}\lambda_0^{\alpha}}{\Gamma(1-\alpha)} \int_{-\infty}^{t} dt'(t-t')^{-\alpha} \frac{d\gamma(t')}{dt'}, \tag{3.25}$$

where the constant, $\mathcal{A} = \dfrac{\tilde{G}}{6\pi a}$. We remark that $\mathcal{A}$ is no longer a constant (typically $\mathcal{A} = \mathcal{A}(t)$) when the concentration effects, such as the bond and entanglement effects, are considered [42]. The right-hand side of Eq. (3.25) represents a fractional integral corresponding to the Caputo formalism [43],

$${}_{-\infty}D_t^{-\beta} f(t) = \frac{1}{\Gamma(\beta)} \int_{-\infty}^{t} \frac{dt'}{(t-t')^{1-\beta}} \frac{df(t')}{dt'}. \tag{3.26}$$

Utilizing Eqs. (3.25, 3.26), we arrive at the basic equation governing stress-strain relation in linear viscoelastic sub-diffusive media,

$$\tau_{xy} = \mathcal{A}\lambda_0^{\alpha} \frac{d^{\alpha-1}}{dt^{\alpha-1}} \frac{d\gamma(t)}{dt} = \mathcal{A}\lambda_0^{\alpha} \frac{d^{\alpha}\gamma}{dt^{\alpha}}, \tag{3.27}$$

including the limiting cases of a purely elastic solid ($\alpha \to 0$ or a Hookean spring) and a purely viscous fluid ($\alpha = 1$ or a dashpot) [44]. Through combinations of springs and dashpots, one arrives at standard linear viscoelastic models, including the Maxwell, Kelvin-Voigt, Zener, Poynting-Thomson and Burgers' model and others [45]. The problem is that the corresponding differential equations have a relatively restricted class of solutions, which are too limited to provide an adequate description for the class of viscoelastic sub-diffusive fluids. To overcome this shortcoming, one can relate the stress and strain through the fractional equation (3.27), which allows a smooth interpolation between a purely elastic behaviour and a purely viscous pattern. In the present analysis, we have selected the Fractional Upper Convected Maxwell

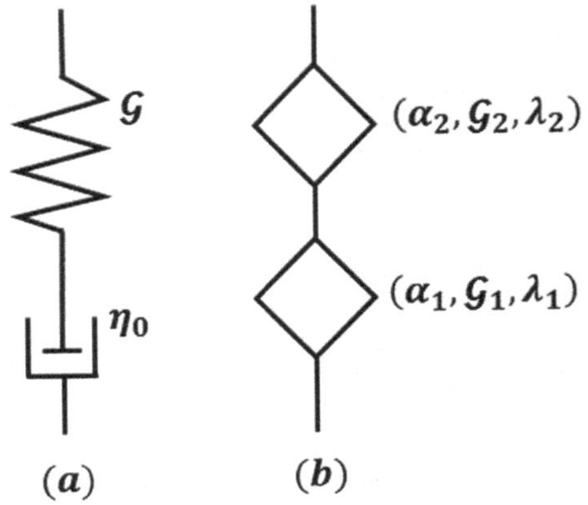

Fig. 3.8 **a** The Maxwell element and **b** its fractional generalization

equation to describe the nonlinear viscoelastic response of the sub-diffusive media, derived next.

Figure 3.8a depicts the standard Maxwell model in which a spring and a dashpot are connected in series [37]. We generalize this model by replacing these elements with their corresponding fractional elements: $(\alpha_i, \mathcal{A}_i, \lambda_i)$, $i = 1, 2$ (Fig. 3.8b). Because of the sequential construction, the stress, τ, is the same for both elements, and their respective stress-strain relations are

$$\gamma_i = \mathcal{A}_i^{-1} \lambda_i^{-\alpha_i} \frac{d^{-\alpha_i} \tau_{xy}}{dt^{-\alpha_i}}, \quad \text{i} = 1, 2, \tag{3.28}$$

where both expressions follow from Eq. (3.27). Due to the construction of the Generalized Maxwell model, we have $\gamma(t) = \gamma_1(t) + \gamma_2(t)$, from which it follows,

$$\tau_{xy} + \frac{\mathcal{A}_1 \lambda_1^{\alpha_1}}{\mathcal{A}_2 \lambda_2^{\alpha_2}} \frac{d^{\alpha_1 - \alpha_2} \tau_{xy}}{dt^{\alpha_1 - \alpha_2}} = \mathcal{A}_1 \lambda_1^{\alpha_1} \frac{d^{\alpha_1} \gamma}{dt^{\alpha_1}}. \tag{3.29}$$

Equation (3.29) can be simplified by setting $\lambda = (\mathcal{A}_1 \lambda_1^{\alpha_1} / \mathcal{A}_2 \lambda_2^{\alpha_2})^{1/(\alpha_1 - \alpha_2)}$ and $E = \mathcal{A}_1 (\lambda_1 / \lambda)^{\alpha_1}$. Without loss of generality in the original spring and dashpot model of Maxwell, as outlined by Schiessel [37], we assume $\alpha_2 = 0$ and $\alpha = \alpha_1 (> 0)$, and arrive at

$$\tau_{xy} + \lambda^{\alpha} \frac{d^{\alpha} \tau_{xy}}{dt^{\alpha}} = \eta_p \frac{d^{\alpha} \gamma}{dt^{\alpha}}, \tag{3.30}$$

where the constant, $\eta_p = E \lambda^{\alpha}$. We can extend equation (3.30) to three dimensions by replacing the elastic stress, τ_{xy}, with the stress tensor, τ, and the derivative, $\frac{d^{\alpha} \gamma}{dt^{\alpha}}$, with the rate of strain tensor, $\mathbf{D} = (\nabla \mathbf{v} + (\nabla \mathbf{v})^T)$ (where the operator $\nabla(\cdot) = \frac{\partial}{\partial \mathbf{x}}(\cdot)$), to arrive at

$$\tau + \lambda^\alpha \frac{d^\alpha \tau}{dt^\alpha} = \eta_p \mathbf{D}, \tag{3.31}$$

using the definition of fractional velocity, $\mathbf{v} = \frac{d^\alpha \mathbf{x}}{dt^\alpha}$ [46], which has a dimension of $\frac{H}{T^\alpha}$ (refer Sect. 3.3.2 for the discussion on non-dimensionalization). Fractional velocities are defined as limits of the difference quotients of a fractional power and they generalize the notion of a local derivative. These derivatives are frequently used, for example, to model instantaneous interactions in Langevin dynamics [46].

Equation (3.31) is the rheological constitutive equation of the Fractional Maxwell model describing the linear viscoelastic media. The simplest way to combine rheological nonlinearity is to replace the (fractional) material time derivative in equation (3.31) with the (fractional) frame invariant, upper-convected time derivative [47], which leads us to the Fractional Upper-Convected Maxwell model, as follows,

$$\tau + \lambda^\alpha \overset{\nabla}{\boldsymbol{\tau}} = \eta_p \mathbf{D}, \tag{3.32}$$

where the fractional upper-convected time derivative of the tensor, τ, is defined as,

$$\overset{\nabla}{\boldsymbol{\tau}} = \frac{\partial^\alpha \tau}{\partial t^\alpha} + \mathbf{v} \cdot \nabla \tau - (\nabla \mathbf{v})^T \tau - \tau \nabla \mathbf{v}. \tag{3.33}$$

The fractional time derivative, $\frac{\partial^\alpha}{\partial t^\alpha}$, in Eqs. (3.32, 3.33) is based on the Caputo definition (3.26).

The continuity and the momentum equations for an incompressible, sub-diffusive flow (consistent with the stress constitutive relation (3.32)) are,

$$\nabla \cdot \mathbf{v} = 0, \qquad \rho \left[\frac{\partial^\alpha \mathbf{v}}{\partial t^\alpha} + \mathbf{v} \cdot \nabla \mathbf{v} \right] = -\nabla p + \eta_s \nabla \cdot \mathbf{D} + \nabla \cdot \tau, \tag{3.34}$$

where ρ is the density, p is the isotropic pressure and $\nabla(\cdot)$ is the gradient operator. At this point we further introduce two new stress tensors which are related to the elastic stress tensor, τ, (a) the elasticity tensor, $\mathbf{A} = \tau/\eta_p$, and (b) the polymer conformation tensor, $\mathcal{C} = \mathbf{I} + We\mathbf{A}$ (more discussion related to this tensor can be found in the next chapter). The parameter, We, is the Weissenberg number, which is related to the time-scale of polymer relaxation, λ^α, to be introduced later. Equations (3.32, 3.34) represent the equations of motion describing the flow-instability of the sub-diffusive viscoelastic fluids.

As a result of the dissipative processes, viscoelastic materials have memory, that is, their actual mechanical response is modulated by the past [48]. The fractional derivative operators account for the complete history to obtain the derivative at an instant. Unlike the classical Maxwell model [6, 49], which accounts for only the elastic (or stored) part of the deformation work, the Fractional Maxwell model accounts for both forms (stored and dissipated) of energy at any time point. Although Mckinley pointed out that the Fractional Maxwell model generally cannot capture polymer shear-thinning [50], the fractional version provides a better fit of the relaxation and creep behaviour for a significantly larger class of viscoelastic materials using fewer parameters than the classical version [48].

3.3.2 Normal Mode Expansion

Next, the linearized stability analyses of the fully developed, planar Poiseuille flow inside an infinitely long channel of length L and width H (i. e., $0 \leq x \leq L$ and $0 \leq y \leq H$ such that $L/H \gg 1$, where x and y are the variables in the flow and the shear gradient direction, respectively) is studied. Figure 3.9 shows the configuration of planar Poiseuille flow. At $y = 0$ and $y = H$, Poiseuille flow is limited within a channel surrounded by solid walls. Mean velocities and stress are not affected by the channel length variable x. However, pressure is affected by x. The mean velocity profile is a function of y and is naturally parabolic [51]. At the channel walls, the mean flow meets the 'no slip' boundary condition. Because the walls are motionless, the fluid velocities at the boundaries are zero. The boundary condition to the perturbation quantities similarly meets the no-slip boundary condition. It is frequently used as a model system for studying fluid dynamics in simple geometries as well as developing theoretical and computational approaches for more complex flow problems. Using the following scales for non-dimensionalizing the governing equations: the height of the channel H for length, the timescale T corresponding to maximum base flow velocity (i. e., $T = (H/\mathcal{U}_0)^{1/\alpha}$) for time and $\rho\mathcal{U}_0^2$ for pressure and stresses, we characterize equations (3.32, 3.34), rephrased as follows,

$$\nabla \cdot \mathbf{v} = 0, \tag{3.35a}$$

$$Re\left[\frac{\partial^\alpha \mathbf{v}}{\partial t^\alpha} + \mathbf{v} \cdot \nabla \mathbf{v}\right] = -\nabla p + \nu \nabla \cdot \mathbf{D} + (1 - \nu)\nabla \cdot \mathbf{A}, \tag{3.35b}$$

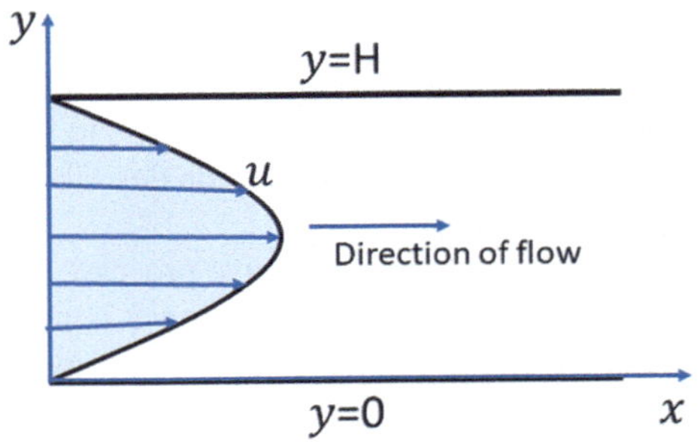

Fig. 3.9 Plane Poiseuille flow

$$\frac{\partial^\alpha \mathbf{A}}{\partial t^\alpha} + \mathbf{v} \cdot \nabla \mathbf{A} - (\nabla \mathbf{v})^T \mathbf{A} - \mathbf{A} \nabla \mathbf{v} = \frac{\mathbf{D} - \mathbf{A}}{We}, \tag{3.35c}$$

using the dimensionless groups, $Re = \dfrac{\rho \mathcal{U}_0 H}{\eta_0}$, $We = \dfrac{\lambda^\alpha \mathcal{U}_0}{H}$ and $\eta_s, \eta_p, \eta_0 (= \eta_s + \eta_p)$ and $\nu (= \eta_s/\eta_0)$ are the solvent viscosity, the polymeric contribution to the shear viscosity, the total viscosity and the viscous contribution to the total viscosity of the fluid, respectively. In equation (3.35), the elastic stress, τ, is related to the polymer stress tensor, $\mathbf{A}$, as $\tau = (1-\nu)\eta_0 \mathbf{A}$. The current analysis deploys fractional derivative of exponentials [43] given as,

$$\frac{d^\alpha (e^{iat})}{dt^\alpha} = (ia)^\alpha e^{iat}. \tag{3.36}$$

Let us denote the mean flow variables with capital letters and with a subscript '0'. We assume that the mean flow is two-dimensional, quasiparallel with its variation entirely in the shear gradient direction. Then, the (non-dimensional) plane Poiseuille flow velocity can be written as follows,

$$\mathbf{U}_0 = \left((y - y^2)\right) \mathbf{e_x}, \tag{3.37}$$

where $\mathbf{e_x}$ is the unit vector along the x-direction. The other mean flow variables satisfying equation (3.35), including the mean pressure, P_0, and the base state elastic stress tensor, $\mathbf{A}_0 = [A_{0_{ij}}]$, is given by,

$$P_0 = -2x - 8We(1-\nu)\left(y - y^2\right), \tag{3.38a}$$

$$A_{0_{11}} = 0, \tag{3.38b}$$

$$A_{0_{12}} = A_{0_{21}} = (1 - 2y), \tag{3.38c}$$

$$A_{0_{22}} = 2We\,(1 - 2y)^2, \tag{3.38d}$$

and whose linearized stability analysis is presented now.

We restrict our stability analysis to the case when the disturbances are two-dimensional. Assuming an independent fate of each wavenumber, k (whose real part is chosen to be positive) and frequency, ω, it is natural to consider disturbances in the form of a normal mode expansion, such that the total velocity, pressure and stress are expressed in terms of their mean values and perturbations amplitudes (denoted by $\mathring{(\cdot)}$), as follows,

$$\mathbf{r} = \mathbf{R}_0 + \epsilon \mathring{r} e^{i(kx - \omega t)}, \tag{3.39}$$

where $\epsilon \ll 1$ and $\mathbf{r} = [\mathbf{v}\; p\; A_{11}\; A_{12}\; A_{22}]^T$, $\mathbf{R}_0 = [\mathbf{U}_0\; P_0\; A_{0_{11}}\; A_{0_{12}}\; A_{0_{22}}]^T$ and $\mathring{r} = [X_0 y(1-y)\; X_1 y(1-y)\; X_2\; X_3\; X_4\; X_5]^T$ represent the total, the mean flow variables and the disturbance amplitudes, respectively. The disturbance amplitude, $\mathring{r}$, is chosen such that it satisfies the no-slip condition on the channel walls. Substituting the solution form (3.39) in equations (3.35a–3.35c) and retaining the $\mathcal{O}(\epsilon)$ terms to arrive at the linearized equation governing the conservation of mass,

$$X_0\left[ik(y-y^2)\right]+X_1\left[1-2y\right]=0, \tag{3.40}$$

the linearized equation describing the conservation of momentum in the x-direction,

$$\begin{aligned}&X_0\left[(y-y^2)\left(Re(-i\omega)^{\alpha}+ikRe(y-y^2)-2\nu(ik)^2\right)+2\nu\right]\\&+X_1\left[Re(1-2y)(y-y^2)-\nu(1-2y)(ik)\right]+ikX_2-ik(1-\nu)X_3=0,\end{aligned} \tag{3.41}$$

and the one governing the conservation of momentum in the y-direction,

$$\begin{aligned}&X_0\left[-ik\nu\,(1-2y)\right]+X_1\left[(y-y^2)\,(Re(-i\omega)^{\alpha}+ikRe\right.\\&\left.(y-y^2)-\nu(ik)^2\right)+4\nu\right]-(1-\nu)ikX_4=0.\end{aligned} \tag{3.42}$$

The linearized equation for the elastic stress component A_{11},

$$\begin{aligned}&X_0\left[\frac{-2}{We}(y-y^2)(ik)\right]+X_1\left[-2ik\,(1-2y)\,(y-y^2)\right]\\&+X_3\left[(-i\omega)^{\alpha}+ik(y-y^2)+\frac{1}{We}\right]=0,\end{aligned} \tag{3.43}$$

for the component A_{12} (or A_{21}),

$$\begin{aligned}&X_0\left[-ik\,(1-2y)\,(y-y^2)-\frac{1}{We}\,(1-2y)\right]\\&+X_1\left[-2(y-y^2)-2ikWe(1-2y)^2(y-y^2)-(1-2y)^2-\frac{1}{We}(y-y^2)(ik)\right]\\&+X_3\left[-(1-2y)\right]+X_4\left[(-i\omega)^{\alpha}+ik(y-y^2)+\frac{1}{We}\right]=0,\end{aligned} \tag{3.44}$$

and for the component A^{22},

$$\begin{aligned}&X_0\left[-2\,(1-2y)^2\right]+X_1\left[-8We\,(1-2y)\,(y-y^2)-4We(1-2y)^3-\frac{2(1-2y)}{We}\right]\\&+X_4\left[-2(1-2y)\right]+X_5\left[(-i\omega)^{\alpha}+ik(y-y^2)+\frac{1}{We}\right]=0.\end{aligned} \tag{3.45}$$

Equations (3.40–3.45) may be written in a matrix-vector format as follows,

$$\begin{bmatrix} ik(y-y^2) & 1-2y & 0 & 0 & 0 & 0 \\ M_1 & M_2 & ik & -ik(1-\nu) & 0 & 0 \\ \nu ik(2y-1) & M_3 & 0 & 0 & -ik(1-\nu) & 0 \\ \frac{2ik}{We}(y^2-y) & 2ik(1-2y)(y^2-y) & 0 & M_4 & 0 & 0 \\ M_5 & M_6 & 0 & -(1-2y) & M_4 & 0 \\ -2(1-2y)^2 & M_7 & 0 & 0 & -2(1-2y) & M_4 \end{bmatrix} \begin{bmatrix} X_0 \\ X_1 \\ X_2 \\ X_3 \\ X_4 \\ X_5 \end{bmatrix} = \begin{bmatrix} 0 \\ 0 \\ 0 \\ 0 \\ 0 \\ 0 \end{bmatrix}, \tag{3.46}$$

where the expressions M_i $(i = 1, \ldots 7)$ utilized above are given as,

$$\begin{aligned}
M_1 &= (y-y^2)\left(Re(-i\omega)^\alpha + ikRe(y-y^2) - 2\nu(ik)^2\right) + 2\nu,\\
M_2 &= Re(1-2y)(y-y^2) - \nu(1-2y)(ik),\\
M_3 &= (y-y^2)\left(Re(-i\omega)^\alpha + ikRe(y-y^2) - \nu(ik)^2\right) + 4\nu,\\
M_4 &= (-i\omega)^\alpha + ik(y-y^2) + \frac{1}{We},\\
M_5 &= -ik\,(1-2y)\,(y-y^2) - \frac{1}{We}\,(1-2y)\,,\\
M_6 &= -2(y-y^2) - 2ikWe(1-2y)^2(y-y^2) - (1-2y)^2 - \frac{1}{We}(y-y^2)(ik),\\
M_7 &= -8We\,(1-2y)\,(y-y^2) - 4We(1-2y)^3 - \frac{2(1-2y)}{We}.
\end{aligned} \tag{3.47}$$

A nontrivial solution for the system (3.46), imposes a zero determinant condition on the coefficient matrix, which leads to the dispersion relation, $D(k,\omega) = 0$, given by,

$$\begin{aligned}
&\frac{1}{We}k^2 M_4\,(2ik\,(-1+\nu)\,(1-2y)\,(-1+y)\,y\,(1+(-2+ikWe\,(1-2y)\\
&(-1+y))\,y) + M_4^2 We\left(\nu\,(1-2y)^2 - M_3\,(-1+y)\,y\right) + M_4\,(-1+\nu)\,We\\
&(M_5 - 2M_5 y + ikM_6\,(-1+y)\,y)) = 0.
\end{aligned} \tag{3.48}$$

3.3.3 Numerical Method

In the ensuing description, we denote real/imaginary components with subscript *r*/*i*, respectively. The zeros of the dispersion relation (3.48) were explored within the complex $k-\omega$ plane inside the region $\omega_r \in [-1700,\ 0.2]$, $\omega_i \in [-5000,\ 100]$, $k_r \in [0,\ 5]$ and $k_i \in [-0.1,\ 0.2]$. For a real wavenumber k, the procedure for finding the most unstable mode (which is the largest positive imaginary component of any root of the dispersion relation or the temporal growth rate, ω^{Temp}, refer Sect. 3.3.5), consists of detecting the admissible saddle points ($\omega \in \mathbb{C}$, $k \in \mathbb{R}$) satisfying the equations [9],

$$D(k, \omega) = 0, \tag{3.49a}$$

$$\frac{\partial \omega_i}{\partial k} = \frac{\partial D/\partial k}{\partial D/\partial \omega_i} = 0, \tag{3.49b}$$

and then (among all the possible roots of equation (3.49)) identifying those roots with the largest positive imaginary component of the frequency. Equation (3.49) is solved using a multivariate Newton-Raphson algorithm (refer to [6, 49] for a detailed outline of this method).

Under the assumption that dispersion relation is a complex analytic function satisfying Cauchy-Riemann relations, the expressions are chosen preferentially to numerically evaluate the derivatives in equation (3.49b). Relaxation of the assumption (of analyticity) does not impact the numerical results, but the evaluation of the complex derivatives in equation (3.49b) becomes cumbersome. The numerical continuation of the temporal growth rate (Sect. 3.3.5) and the absolute growth rate curves (Sect. 3.3.6) were realized within the range $Re \in [10^{-6}, 10^{2}]$, $We \in [0, 10^{3}]$ and at two specific values of $\nu = 0.05$ (the elastic stress-dominated case) and $\nu = 0.3$ (the viscous stress-dominated case), using a discrete step-size of $\triangle Re = 10^{-6}$ and $\triangle We = 10^{-4}$, respectively. While the (non-dimensional) physical domain spans within the range, $x \in (-\infty, \infty)$; $y \in [0, 1]$, the temporal growth rate (ω_i^{Temp}) and the absolute growth rate (ω_i^{Cusp}) of the perturbations are probed at four discrete, transverse spatial locations of the advancing interface: $y = 0.2, 0.9, 0.7, 0.5$. While the former two values of y are chosen qualitatively to probe the near-wall effects, the last value is selected to understand the development of the centerline instability.

3.3.4 Model Validation

The model and the numerical method outlined above is validated by reproducing the neutral stability curves for a plane Poiseuille of a classical Oldroyd-B fluid, as investigated by Atalik [52] ($\alpha = 1.0$ or the red curves in Fig. 3.10, in comparison with Fig. 6 in [52]. The 'dots' on the curves are the only parameter values where the simulation occurred). The neutral stability curves for two fractional orders ($\alpha = 0.99$ (blue curves) and $\alpha = 0.95$ (green curves)) are also shown for comparison. The locus of neutrally stable points is found after selecting $\omega_i = 0$ in the dispersion relation (3.48) and solving for the unknowns (ω_r, k), at fixed values of Reynolds and elasticity number ($E = We/Re$).

Two conclusions can be deduced from Fig. 3.10. First, notice that the minimum value of the critical Reynolds number predicting a temporal instability increases, with increasing viscosity ratio, both for the classical case (a result identical to the

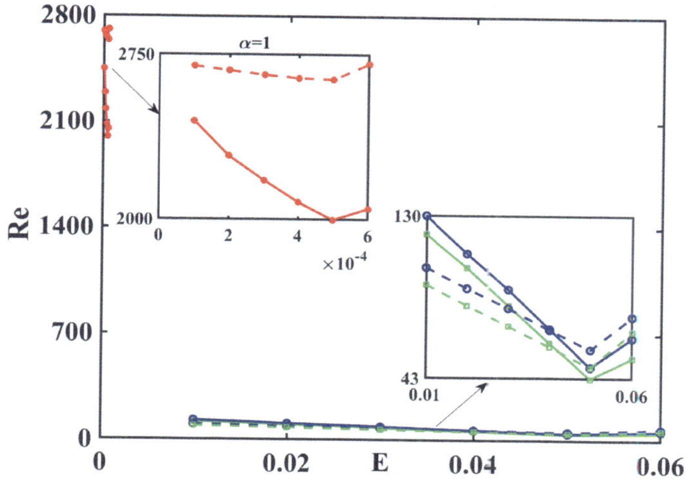

Fig. 3.10 Neutral stability curves at the centerline ($y = 0$) for plane Poiseuille viscoelastic flow of a classical fluid (red curves, source: Fig. 6 in [52]) versus sub-diffusive fluid at $\alpha = 0.99$ (blue curves), $\alpha = 0.95$ (green curves) and at viscosity ratio, $\nu = 0.01$ (solid curves), $\nu = 0.9$ (dashed curves), projected onto the $Re - E$ plane

one predicted by Atalik [52]) as well as the sub-diffusive case. Second, observe that this minimum value of Re is significantly lower and appears at significantly larger values of E, for the sub-diffusive fluid. These two observations indicate that the transition to instability is primarily driven by elasticity (rather than fluid inertia) for sub-diffu2sive fluids. A more detailed outlook of the influence of elasticity is acquired by examining the temporal growth rates, described next.

3.3.5 *Temporal Stability Analysis for Poiseuille Flows*

First, we explore the linear stability of the model (3.46) by exclusively assigning ω to be a complex number. In earlier studies on wall-bounded viscoelastic flows, elasticity (characterized by the parameter, We) was found to have a destabilizing effect (for example, see [34] and the references within). In this section, we partially extend some of these ideas for the sub-diffusive, two-dimensional Poiseuille flow within a selected range of parameters, Re, We, ν and especially for the case of the coarse-grained Rouse chain melts and the coarse-grained Zimm chain solution, which corresponds to the fractional order derivatives, $\alpha = 1/2, 2/3$, respectively. Both these models have been extensively studied at the molecular level by the polymer science community [40], mainly for their simplicity and wide range of applicability. We briefly digress to describe the rheological significance of these two classes of polymer melts.

3.3.5.1 Coarse-Grained Rouse Chain Melts: $\alpha = 1/2$

The coarse-grained Rouse model [35] (referred simply as Rouse model from hereon) predicts that the viscoelastic properties of the polymer chain can be described by a Generalized Maxwell model, where the elasticity is governed by a single relaxation time, which is independent of the number of Maxwell elements (or the so-called 'submolecules'). In polymer melts, the Rouse model correctly predicts the long-time diffusion only for chains shorter than the entanglement length. For long chains with noticeable entanglement, the Rouse model holds only up to a characteristic crossover time. For longer times the chain can only move within a tube formed by the surrounding chains, a motion which is usually approximated by the reputation model [53].

3.3.5.2 Coarse-Grained Zimm Chain Solution: $\alpha = 2/3$

The coarse-grained Zimm's model [36] (referred simply as Zimm model from hereon) predicts the ('shear rate and polymer concentration independent') viscosity of the polymer solution by calculating the hydrodynamic interaction of flexible polymers (an idea which was originally proposed by Kirkwood [54]) by approximating the chains using a bead-spring setup. While the Rouse model applies to polymer melts, the Zimm model applies to polymer in solution where the hydrodynamic interaction is not screened. In a solution, both Rouse and Zimm's model predicts the diffusivity in the order of $D \sim 1/N^{\alpha}$ (where N is related to the molecular weight of the polymer chains), which is consistent with experiments [53].

3.3.5.3 Numerical Results

Next, we resume our temporal instability investigation. Figure 3.11 present the variation of the most unstable mode versus Re, and at fixed We and ν for viscoelastic Poiseuille flow. Observe that the elastic stress-dominated case (or $\nu = 0.05$ case) is temporally more unstable (i. e., compare the maximum 'y' value on the ordinate axis of the figures on the left column versus those on the right column in Fig. 3.11). Also, observe especially for the Zimm's case, that not only the peak of the most unstable mode increases but also the range of Reynolds number exhibiting temporal instability increases with increasing values of We (i. e., notice the dashed green, blue and the red curves in Fig. 3.11). Also, analogous with the traditional (or integer order) viscoelastic channel flows, we find that for intermediate values of Reynolds number (or $1 \leq Re \leq 55$), elasticity is destabilizing (notice, from the dashed curves in figure insets in Fig. 3.11, that the most unstable mode is larger for larger values of We). These observations lead us to conclude that elasticity has a destabilizing impact within the intermediate range of Re. This destabilization mechanism is the result of a complex interaction between the inertial forces (typically operative at

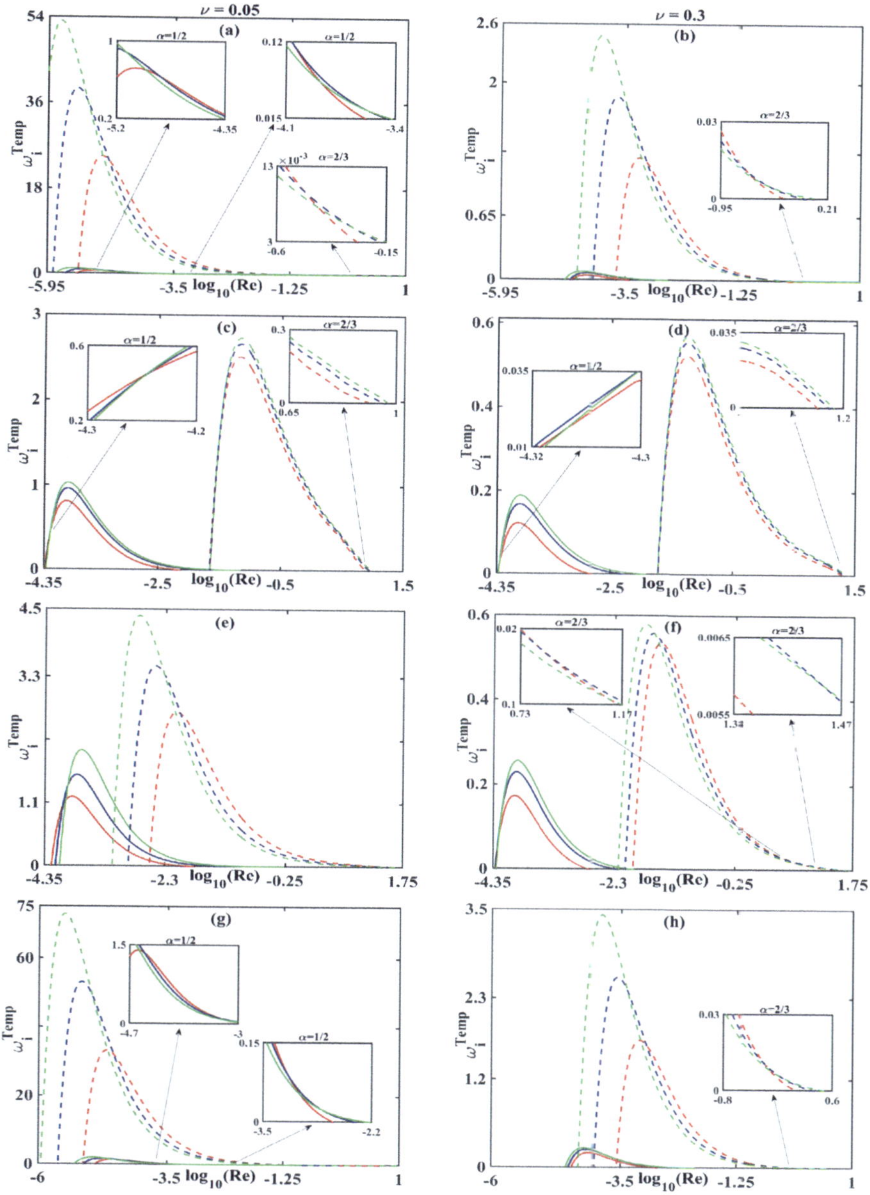

Fig. 3.11 Most unstable mode, ω_i^{Temp} versus Reynolds number for the Rouse model (solid curves) and for the Zimm's model (dashed curves), evaluated at $We = 15.0, 25.0, 35.0$ (red, blue and green curves, respectively), viscosity ratios, $\nu = 0.05$ (left column) and $\nu = 0.3$ (right column) and at transverse spatial locations: **a**, **b** $y = 0.2$, **c**, **d** $y = 0.5$, **e**, **f** $y = 0.7$ and **g**, **h** $y = 0.9$

larger Reynolds number) and the normal stress anisotropy through elasticity (proportional to We) and can be explained via an energy formalism: the stretching of the polymers with increasing elasticity brings about a normal stress anisotropy, leading to an elastically loaded fluid, that is, when the polymers stretch, elastic energy is stored in the sheared fluid. This energy is transferred and released after the fluid element has been advected to other regions where the shear-induced stretching forces are smaller [47]. However, for sufficiently larger values of Re (or $Re > 55$), we find the emergence of the temporally stable state. The appearance of the temporally stable state at high fluid inertia is a hallmark of sub-diffusive flows, and the details of the same are elaborated in Sect. 3.3.6.

Regarding the near-wall effects, notice that Zimm's case is more unstable near the wall (i. e., comparing the maximum 'y' value on the ordinate axis for $y = 0.2, 0.9$, Fig. 3.11a, b and g, h respectively) in comparison with the corresponding instability on channel centerline ($y = 0.5$ case, Fig. 3.11c, d). Further, within the intermediate range of Reynolds number (i. e., $1 \leq Re \leq 55$), elasticity is destabilizing near the walls (e. g., see the dashed curves in figure insets in Fig. 3.11a, b, g, h). All these near-wall effects can be understood via a mechanism similar to the one proposed by Rabaud [55] for wall-bounded Newtonian flows: the boundary effects induce a local perturbation on the advancing interface which (when coupled with elasticity) destabilizes the flow.

Finally, we find that the order of the fractional derivative, α has a strong correlation with the temporal stability of the channel flows. Comparing the values of ω_i^{Temp}, we deduce that the Zimm's model is temporally more unstable than the Rouse case. In a series of in silico studies, an investigation of the most unstable mode within the range, $\alpha \in [0.5\ 1.0]$, reveals: (a) the most unstable mode decreases with decreasing order of the fractional derivative, α, (b) the peak of the most unstable mode shifts to lower values of Re with decreasing values of α, and (c) in particular, the peak of the most unstable mode, for the Rouse model (i. e., the solid curves in Fig. 3.11), precipitates towards the limit $Re \rightarrow 0$. In other words, the transition pathway to flow turbulence in the Rouse polymer flows is characterized via elastic turbulence (appearing at vanishingly low values of Re and at moderate to high values of We) [56]. To summarize, as α decreases, the nature of the transition pathway to flow turbulence changes from that of the elastoinertial turbulence (characterized by moderate values of Re and We) to elastic turbulence.

We recapitulate the interplay of the inertial forces (characterized by the parameter Re), the elastic forces (represented by the parameter We) as well as the boundary effects and the order of the fractional derivative on the progression of the temporal instability (exemplified by the most unstable mode) of the viscoelastic sub-diffusive channel flows as follows: elasticity combined with reasonable fluid inertia has a destabilizing impact on the evolving flow front. The finite boundary is shown to have a destabilizing influence. Finally, the order of the sub-diffusive timescale (alternatively, the order of the fractional derivative) impacts the nature of the transition pathway to turbulence (if any). In the next section, we outline a deeper characterization of these instabilities via spatiotemporal analysis.

3.3.6 Spatiotemporal Stability Analysis for Poiseuille Flows

The spatiotemporal analysis is typically relevant when one introduces an impulse excitation locally in a flow and observes how that disturbance evolves in time [9]. More significantly, we evaluate the absolute growth rate (or ω_i^{Cusp}, details on computing these points are elaborated in Sect. 3.3.3) to identify the region of absolute instability. However, evanescent modes are also encountered in our analysis [6]. These modes do not merely depend on the sign of the absolute growth rate and have to be found via the sufficient conditions proposed by Briggs [30] (refer Sect. 3.3.3). Evanescent modes are briefly referred in the description of the phase diagrams (Fig. 3.13).

Figure 3.12 represents the absolute growth rate curves versus Re, at three fixed values of Weissenberg number, $We = 15.0, 25.0, 35.0$, and at $\nu = 0.05$ (the elastic stress-dominated case) and $\nu = 0.3$ (the viscous stress-dominated case). For the selected values of We and for the elastic stress-dominated case, we find that the Rouse model exhibits a transition from absolute instability (or the parameter values where $\omega_i^{\text{Cusp}} > 0$) towards temporal stability (or the range of parameters where no curve is drawn in the plots for absolute growth rate curves) at a critical value of Reynolds number, $Re_c < 10^{-3}$, at $y = 0.2$ (Fig. 3.12a). This critical Reynolds number increases as one moves closer to the upper plate and along the transverse spatial location, y (i. e., compare the Re_c values from the inset in Fig. 3.12a, c, e, g). For the viscous stress-dominated case, the Rouse model indicates the following transition with increasing values of Re: convective instability (or the parameter values where $\omega_i^{\text{Cusp}} < 0$)→ absolute instability→ temporal stability. Again, the critical value of the Reynolds number at these transition points increases as one progressively moves towards the upper plate (refer to Fig. 3.12b, d, f, h). In contrast, the Zimm's model highlights a direct transition from absolute instability towards temporal stability, in the range of low to moderate values of Re and for both the elastic as well as the viscous stress-dominated case. However, both of these models reveal temporal stability in the limit of vanishingly small Reynolds number (or in the strongly elastic limit).

We remark that while some observations listed above follow from the well-established mechanisms seen in classical (or integer order) viscoelastic flows, namely the lack of symmetry in the flow-instability transition across the centerline (due to the anisotropy of the elastic stresses) as well as the appearance of absolute/convective instabilities at intermediate values of Re (generated due to the instability via the polymer elasticity), other observations are relatively novel, specifically the flow-induced (temporal) stabilization at higher values of Re.

Next, we classify the nature of these instabilities by computing the boundaries of the temporally stable regions (**S**), evanescent modes (**E**), the convectively unstable (**C**) and the absolutely unstable regions (**A**) within a selected range of the flow-elasticity-viscosity parameter space, i.e., $Re \in [10^{-6}, 100]$, $We \in [0, 10^3]$ and $\nu \in \{0.05, 0.30\}$. While convective instability grows in amplitude as it is swept along by the flow, absolute instability occurs at fixed spatial locations, leading to surface transitions (or pinch-off) of the advancing interface [9]. The flow stability phase

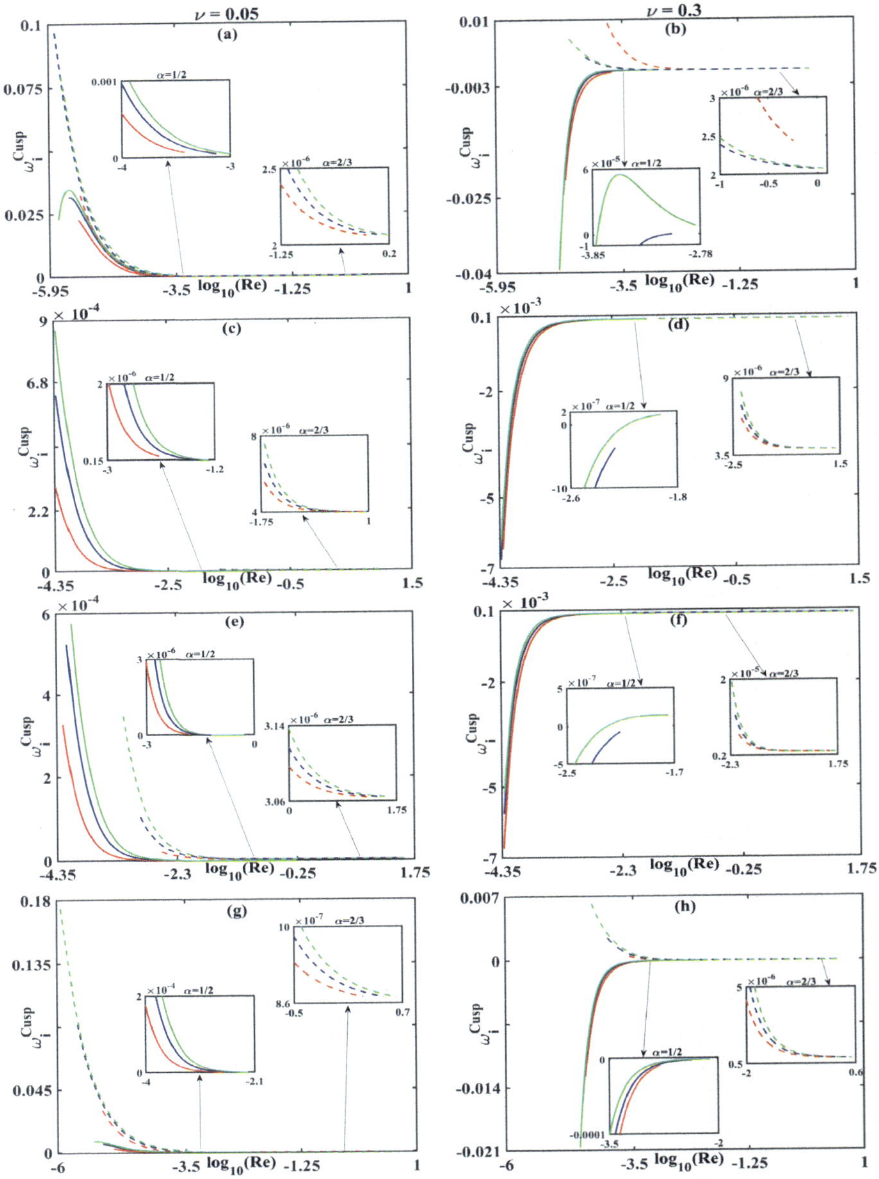

Fig. 3.12 Absolute growth rate, ω_i^{Cusp} versus Reynolds number for the Rouse model (solid curves) and for the Zimm's model (dashed curves), evaluated at $We = 15.0, 25.0, 35.0$ (red, blue and green curves, respectively), viscosity ratios, $\nu = 0.05$ (left column) and $\nu = 0.3$ (right column) and at transverse spatial locations: **a**, **b** $y = 0.2$, **c**, **d** $y = 0.5$, **e**, **f** $y = 0.7$ and **g**, **h** $y = 0.9$

diagram, projected onto the $Re - We$ parameter space (Fig. 3.13) reveal the presence of absolutely unstable and convectively unstable region at low to moderate values of Re and We ($Re \in [10^{-5}, 10]$ and $We \in [3, 350]$ for absolute instability, and $Re \in [10^{-5}, 10^{-2}]$ and $We \in [50, 150]$ for convective instability, respectively), as result of a complex tug-of-war between the inertial forces (proportional to Re) and the normal stress anisotropy through elasticity (proportional to We). To summarize, the parameter regions susceptible to topological transitions (or the parameter space which indicates absolute instability) in sub-diffusive channel flows, are those driven by moderate inertia coupled with moderate to high elasticity.

An exceptional feature in the phase diagrams (3.13) is the presence of temporal stability at high inertia (i. e., $Re \geq 55$), whose origin is described in detail in the next chapter. Finally, we remark that these in vitro studies not only corroborate our numerical outcome, especially establishing the emergence of the temporally stable region at high inertia but also highlight the potential of fractional calculus in effectively capturing the flow-instability transition in sub-diffusive flows.

3.4 Conclusions

This chapter addresses the temporal and the spatiotemporal linear stability analyses of viscoelastic, sub-diffusive, plane Poiseuille flow in the limit of low to moderate Reynolds number and moderate to high Weissenberg number. Sections 3.3.1, 3.3.2 and 3.3.3 presented the viscoelastic, sub-diffusive channel flow model, the elements of linear stability analysis and the numerical method needed to solve the resulting dispersion relation. Section 3.3.4 validated the model for the classical planar Poiseuille flow obeying the Oldroyd-B stress constitutive equation [52]. The temporal stability analysis in Sect. 3.3.5 indicates that with decreasing order of the fractional derivative: (a) the most unstable mode decreases, (b) the peak of the most unstable mode shifts to lower values of Re, and (c) in particular, the peak of the most unstable mode, for the Rouse model converges towards the limit $Re \rightarrow 0$. The spatiotemporal phase diagram in Sect. 3.3.6 indicates an abnormal region of temporal stability at high fluid inertia coupled with high elasticity, due to the presence of a non-homogeneous environment with hindered flow. Although we have shown how the exponent in the sub-diffusive power-law scaling of the mean square displacement of the tracer particle in the microscale is related to the fractional order of the corresponding nonlinear stress constitutive equations in the continuum, the arguments presented herein are 'phenomenological' in nature.

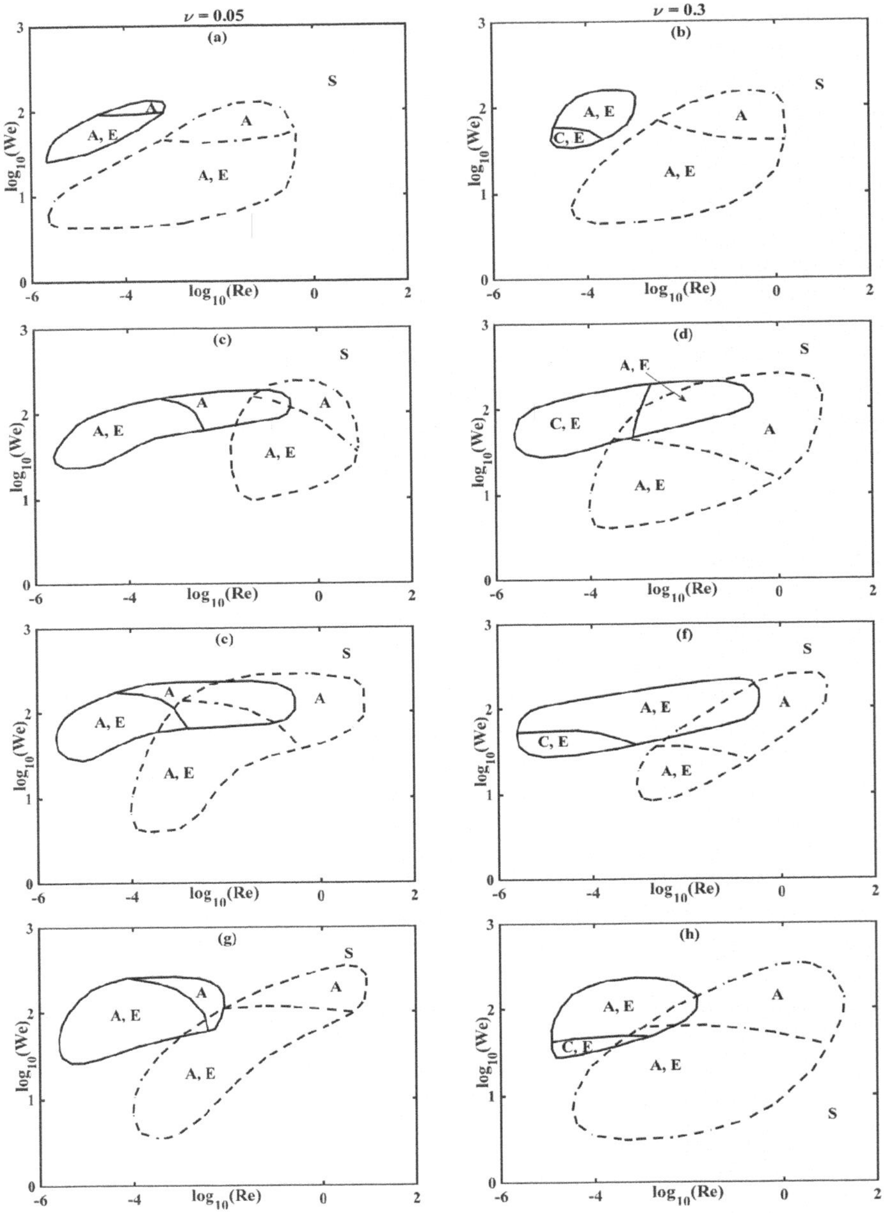

Fig. 3.13 Viscoelastic sub-diffusive stability phase diagram in the $Re - We$ parametric space for the Rouse model (solid curves) and for the Zimm's model (dashed curves), evaluated at transverse spatial locations: **a**, **b** $y = 0.2$, **c**, **d** $y = 0.5$, **e**, **f** $y = 0.7$ and **g**, **h** $y = 0.9$ and at fixed values of viscosity ratios, $\nu = 0.05$ (left column) and $\nu = 0.3$ (right column)

3.5 Problems

1. Consider a two-dimensional plane Couette flow of a viscoelastic fluid confined between two infinite parallel plates located at $y = \pm h$. Let the top (located at $y = h$) and bottom (located at $y = -h$) plates be moving with a constant speed U in the positive x- and negative x-directions, respectively. Assuming the upper-convected Maxwell model for viscoelasticity,
   a. Write down the governing equations and make them dimensionless them by choosing appropriate scaling,
   b. Solve the governing equations for the steady-state velocity profile, $u(y)$.
2. Perform a linear stability analysis of the base state of the plane Couette flow obtained in Problem 1 and derive the associated Orr–Sommerfeld equation.
3. Consider a two-dimensional plane Poiseuille flow of a viscoelastic fluid confined between two infinite parallel plates located at $y = \pm h$. Given that the plane Poiseuille flow is driven by a constant longitudinal pressure gradient. Assuming the upper-convected Maxwell model for viscoelasticity,
   a. Write down the governing equations and make them dimensionless by choosing appropriate scaling,
   b. Solve the governing equations for the steady-state velocity profile, $u(y)$.
4. Perform a linear stability analysis of the base state of the plane Poiseuille flow obtained in Problem 3 and derive the associated Orr–Sommerfeld equation.

References

1. Alexander NM, Saarloos WV (2007) An introductory essay on subcritical instabilities and the transition to turbulence in viscoelastic parallel shear flows. Phys Rep 447(3):112–143
2. Ho CM, Huerre P (1984) Perturbed free shear layers. Annu Rev Fluid Mech 16:365–424
3. Azaiez J, Homsy GM (1994) Linear stability of free shear flow of viscoelastic liquids. J Fluid Mech 268:37–69
4. Schmid P, Henningson D (2001) Stability and transition in shear flows. Springer-Verlag, New York Inc, p 142
5. Adelman SA (1976) Fokker-Planck equations for simple non-Markovian systems. J Chem Phys 64(1):124–130
6. Bansal D, Ghosh D, Sircar S (2021) Spatiotemporal linear stability of viscoelastic free shear flows: nonaffine response regime. Phys Fluids 33:054106
7. Crighton DG, Gaster M (1976) Stability of slowly-diverging jet flow. J Fluid Mech 77:397–424
8. Gaster M, Kit E, Wygnanski I (1985) Large scale structures in a forced turbulent mixing layer. J Fluid Mech 150:23–29
9. Huerre P, Monkewitz PA (1990) Local and global instabilities in spatially developing flows. Ann Rev Fluid Mech 22:473–537
10. Briggs RJ (1964) Electron-stream interaction with plasmas. MIT Press
11. Thacker WC (1976) Spatial growth of gulf stream meanders. Geophys Fluid Dyn 7(27):1–95

12. Merkine LO (1977) Convective and absolute instability of baroclinic eddies. Geophys Astrophys Fluid Dyn 9:29–57
13. Twiss RQ (1952) Propagation in electronion streams. Phys Rev 88:1392–1407
14. Landau LD, Lifshitz EM (1959) Fluid mechanics. Permagon Press
15. Sturrock PA (1958) Kinematics of growing waves. Phys Rev 112:1488–1503
16. Lifshitz EM, Pitaevskii LP (1981) Physical kinetics. Pergamon, London
17. Bers A (1983) Space-time evolution of plasma instabilities-absolute and convective. Handb Plasma Phys 1
18. Clemmow PC, Dougherty JP (1969) Electronics of particles and plasmas. Addison-Wesle, Reading, Mass
19. Pierce AD (1981) Acoustics: an introduction to its physical principles and applications. McGraw-Hill
20. Koch W (1986) Direct resonance in Orr-Sommerfeld problems. Acta Mech 59:11–29
21. Akylas TR (1982) A nonlinear theory for the generation of water waves by wind. Stud Appl Math 67:1–24
22. Gustavsson LH, Hultgren LS (1980) A resonance mechanism in plane Couette flow. J Fluid Mech 98:149–159
23. Gustavsson LH (1981) Resonant growth of three-dimensional disturbances in plane Poiseuille flow. J Fluid Mech 112:253–264
24. Hultgren LS, Gustavsson LH (1981) Algebraic growth of disturbances in a laminar boundary layer. Phys Fluids 24:1000–1004
25. Akylas TR, Benney DJ (1980) Direct resonance in nonlinear wave systems. Stud Appl Math 63:209–226
26. Akylas TR, Benney DJ (1982) The evolution of waves near direct-resonance conditions. Stud Appl Math 67:107–123
27. Aranha JA, Yue DKP, Mei CC (1982) Nonlinear waves near a cut off frequency in an acoustic duct- a numerical study. J Fluid Mech 121:465–485
28. Cremer L (1953) Theorie der Luftschall-Di mpfung im l echteckkanal mit schluckender Wand and das sich dabei ergebende hSchste DS mpfungsmag. Acustica 8:249–263
29. Benny DJ, Gustavsson LH (1981) A new mechanism for linear and nonlinear hydrodynamic instability. Stud Appl Math 64:185–209
30. Kupfer K, Bers A, Ram AK (1987) The cusp map in the complex-frequency plane for absolute instability. Phys Fluids 30(10):3075–3082
31. Lingwood RJ (1997) Absolute instability of the Ekman layer and related rotating flows. J Fluid Mech 331:405–428
32. Yeo KS, Khoo BC, Zhao HZ (1996) The absolute instability of boundary-layer flow over viscoelastic walls. Theor Comput Fluid Dyn 8(4):237–252
33. Mason TG, Weitz DA (1995) Optical measurements of frequency-dependent linear viscoelastic moduli of complex fluids. Phys Rev Lett 74(7):1250–1253
34. Khalid M, Chaudhary I, Garg P, Shankar V, Subramanian G (2021) The centre-mode instability of viscoelastic plane Poiseuille flow. J Fluid Mech 915(A43)
35. Rouse PE (1953) A theory of the linear viscoelastic properties of dilute solutions of coiling polymers. J Chem Phys 21(7):1272–1280
36. Zimm BH (1956) Dynamics of polymer molecules in dilute solution: viscoelasticity, flow birefringence and dielectric loss. J Chem Phys 24(2):269–278
37. Schiessel H, Metzler R, Blumen A, Nonnenmacher TF (1995) Generalized viscoelastic models: their fractional equations with solutions. J Phys A 28(23):6567–6584
38. Brader JM (2010) Nonlinear rheology of colloidal dispersions. J Phys Condens Matt 22(363101)
39. Mason TG, Gang H, Weitz DA (1996) Rheology of complex fluids measured by dynamic light scattering. J Mol Struct 383:81–90
40. McKinley S, Yao L, Forest MG (2009) Transient anomalous diffusion of tracer particles in soft matter. J Rheol 53(6):1487–1506

41. Sircar S, Roberts AJ (2016) Surface deformation and shear flow in ligand mediated cell adhesion. J Math Biol 73(4):1035–1052
42. Kremer K, Grest GS (1990) Dynamics of entangled linear polymer melts: a molecular-dynamics simulation. J Chem Phys 92(8):5057–5086
43. Glockle WG, Nonnenmacher TF (1994) Fractional relaxation and the time-temperature superposition principle. Rheo Acta 33:337–343
44. Scott-Blair GW (1944) Analytical and integrative aspects of the stress-strain-time problem. J Sci Instr 21(5):80–84
45. Scott-Blair GW (1947) The role of psychophysics in rheology. J Coll Sci 2(1):21–32
46. Prodanov D (2018) Fractional velocity as a tool for the study of non-linear problems. Fractal Fract 2(1)
47. Spagnolie SE (2015) Complex fluids in biological systems: experiment, theory, and computation. Springer
48. Jiménez AH, Santiago JH, García AM, Gonzáles JS (2002) Relaxation modulus in PMMA and PTFE fitting by fractional Maxwell model. Poly Test 21:325–331
49. Sircar S, Bansal D (2019) Spatiotemporal linear stability of viscoelastic free shear flows: dilute regime. Phys Fluids 31(8):084104
50. Jaishankar A, McKinley GH (2014) A fractional K-BKZ constitutive formulation for describing the nonlinear rheology of multiscale complex fluids. J Rheo 58(6):1751–1788
51. Thomas LH (1953) The stability of plane Poiseuille flow. Phys Rev 91:780–783
52. Atalik K, Keunings R (2002) Non-linear temporal stability analysis of viscoelastic plane channel flows using a fully-spectral method. J Non-Newt Fluid Mech 102:299–319
53. Doi M, Edwards SF (1986) The theory of polymer dynamics. Clarendon Press, Oxford University, New York
54. Kirkwood JG (1954) The general theory of irreversible processes in solutions of macromolecules. J Poly Sci 12(1):1–14
55. Rabaud M, Couder Y, Gerard N (1988) Dynamics and stability of anomalous Saffman-Taylor fingers. Phys Rev A 37:935–947
56. Larson RG (2000) Turbulence without inertia. Nature 405:27–28

Chapter 4
Macrostructure Quantification Via a Riemannian Metric

Abstract This chapter presents a theory to quantify the formation of spatiotemporal macrostructures (or the non-homogeneous regions of high viscosity at moderate to high fluid inertia) for viscoelastic sub-diffusive flows, by introducing a mathematically consistent decomposition of the polymer conformation tensor, into the so-called Structure Tensor. Our approach bypasses an inherent problem in the standard arithmetic decomposition, namely, the fluctuating conformation tensor fields may not be positive definite and hence, do not retain their physical meaning. Using well-established results in matrix analysis, the space of positive definite matrices is transformed into a Riemannian manifold by defining and constructing a geodesic via the inner product on its tangent space. This geodesic is utilized to define three scalar invariants of the Structure Tensor, which do not suffer from the caveats of the regular invariants (such as trace and determinant) of the polymer conformation tensor. We conclude by considering the problem of formulating perturbative expansions of the structure tensor using this geodesic, such that a constraint on the maximum time, during which the evolution of the perturbative solution can be well approximated by linear theory along the Euclidean manifold is found.

4.1 Introduction

As we saw in Chap. 1, a notably abnormal feature in viscoelastic sub-diffusive flows is the presence of temporally stable, non-homogeneous regions of high viscosity at moderate to high fluid inertia (or the so-called spatiotemporal macrostructures). For example, Riley [1] reported an elasticity-induced flow stabilization of viscoelastic fluids coated over compliant surfaces at a fairly high Reynolds number ($Re \sim 4000$). In a separate study involving ethanol gel fuels, elastic stabilization at a high shear rate was attributed to an abnormally high second normal stress difference [2]. Viscoelastic flow stabilization at higher values of Re, in tapered microchannels, was explained due to the presence of wall effects [3]. In another in vitro study, biofilm deacidification created a non-homogeneous environment for molecular diffusion, leading to a 'sub-diffusive effect' with hindered flow rates [4]. In summary, while the in silico studies of the classical channel flows indicate the appearance of temporal instability for

H. Wilson et al., *Viscoelastic Subdiffusive Flows*, Fluid Mechanics and Its Applications 138, https://doi.org/10.1007/978-981-96-0135-6_4

Reynolds number as low as $Re \sim 50$ [5], temporal stability and 'structure formation' for viscoelastic sub-diffusive flows is only recognized in experimental realizations, until now.

Previous approaches to analyze the dynamics of structure formation in dilute solutions have been to utilize the statistics of polymer forces [6] and torques [7]. However, a more appropriate quantity to probe the polymer deformation history is the conformation tensor, $\mathcal{C}$, a second-order positive definite tensor, which is obtained by averaging, over all molecular realizations, the dyad formed by the polymer end-to-end vector [8]. The trace of $\mathcal{C}$ (denoted from here onwards as tr $\mathcal{C}$) is commonly used in literature to analyze $\mathcal{C}$ since (i) it is equal to the sum of its principal stretches and is, therefore, a measure of the polymer deformation [9], (ii) it is proportional to the elastic energy in purely Hookean constitutive models of the polymers [10]. However, tr $\mathcal{C}$ is not a sufficiently complete descriptor of polymer deformation. For example, Berris [10] found that the mean of tr $\mathcal{C}$ can increase with increasing elasticity without a commensurate effect on the mean velocity profile. This behaviour arises because the mean stress deficit is not a function of any of the normal components of $\mathcal{C}$. This example highlights the importance of simultaneously considering all of the components of $\mathcal{C}$ in order to arrive at a complete picture of the polymer deformation and its effect on the velocity field. The fluctuating conformation tensor, $\mathcal{C}'$ (obtained by subtracting the mean conformation tensor, $\overline{\mathcal{C}}$ from instantaneous tensor $\mathcal{C}$) and its moments, provide one method to obtain relevant higher-order statistical descriptions of $\mathcal{C}$. However, this fluctuating tensor is not guaranteed to be physically realizable since (i) whenever tr $\mathcal{C}' \leq 0$, this implies negative material deformation and this tensor loses positive-definiteness, and (ii) equally probable states of contraction (tr $\mathcal{C} \in (0, 1)$) and expansion (tr $\mathcal{C} \in (1, \infty)$) would be described by fluctuations with very different magnitudes. A more appealing way to evaluate fluctuations in $\mathcal{C}$ is to use $\log \mathcal{C}$ because the logarithm of a positive definite matrix is a symmetric matrix, and the set of symmetric matrices form a vector space [11]. While $\log \mathcal{C}$ has been an object of interest in some studies of viscoelastic flows [12], two additional difficulties arise in using $\log \mathcal{C}$. First, the mean value of $\mathcal{C}$, or $\overline{\mathcal{C}}$, is not equal to $e^{\log \overline{\mathcal{C}}}$ implying that the effect of the polymer stress on the mean momentum balance requires all statistical moments of $\log \mathcal{C}$, even when the polymer stress is a linear function of $\mathcal{C}$. A second difficulty is that, in general, $e^{\log \overline{\mathcal{C}} + (\log \mathcal{C}')} \neq e^{\log \overline{\mathcal{C}}} \cdot e^{(\log \mathcal{C}')}$, which implies that there is no way to associate $\log \mathcal{C}'$ with a physical polymer deformation. Thus, this chapter is dedicated to the development of an alternate Structure Tensor from the polymer conformation tensor $\mathcal{C}$ as well as a formal way to visualize this new tensor for sub-diffusive flows.

Although the mathematical results outlined in this chapter have a well-established background in advanced matrix analysis textbooks [11, 13], to the author's best knowledge, they have not been used to evaluate the hydrodynamics of sub-diffusive flows. Here, we aim to (i) derive an appropriate tensor (or the so-called 'Structure Tensor') that describes the polymer deformation in a physically realizable manner, (ii) derive appropriate scalar measures associated with the Structure Tensor, and (iii) corroborate our theory developed in aim-(i) and (ii) through regular perturbation analysis. The chapter is organized as follows. Section 4.2 provides a basic framework on Riemannian manifolds, on the basis of which we describe our mathematical model in Sect. 4.3. Equations describing the dynamics of the structure tensor are presented in Sect. 4.4. Section 4.5 outlines the main result, namely, the description of three scalar invariants of the Structure Tensor via the development of a geodesic on the Riemannian manifold. The weakly nonlinear perturbation analysis as well as a detailed derivation of the linear perturbed solution, is outlined in Sect. 4.6. The conclusions follow in Sect. 4.7.

4.2 Basics of Riemannian Manifolds

The purpose of this section is to introduce the fundamentals of *Riemannian manifolds* or smooth manifolds equipped with Riemannian metrics (which are smoothly varying choices of inner products on tangent spaces). These metrics allow one to measure geometric quantities such as distances and angles. Further, we discuss some key results related to the completeness of these manifolds; in particular, we will present the statement and the proof of the Hopf-Rinow Theorem [13, 15, 16]. We begin with the definitions of Riemannian metrics and isometries (or the distance-preserving maps).

Definition 4.1 (*Riemannian metric*) Let M be a smooth manifold. A *Riemannian metric* on M is a smooth covariant 2-tensor field $g \in \mathcal{T}^2(M)$ whose value g_p at each $p \in M$ is an inner product on T_pM (the tangent space).

Alternatively, the Riemannian metric, g, is a symmetric 2-tensor field that is positive definite in the sense that $g_p(v, v) \geq 0$ for each $p \in M$ and for each $v \in T_pM$, with the equality holding if and only if $v = 0$. The pair (M, g) is said to be a Riemannian manifold. We often use an angle-bracket notation for the inner product. For example,

$$g_p(v, w) = \langle v, w \rangle_g,$$

where $v, w \in T_pM$. Using this inner product, we can define lengths of tangent vectors, angles between nonzero tangent vectors and orthogonality of tangent vectors. For example, the length of a vector $v \in T_pM$ is denoted by $|v|_g = \langle v, v \rangle^{1/2}$. In our subsequent discussion, we will drop the subscript 'g' from the inner product in the definition above. For example, note that the *Euclidean metric* is a Riemannian metric on $\mathbb{R}^n$ whose value at each point $x \in \mathbb{R}^n$ is the usual dot product on its tangent space $T_x\mathbb{R}^n$, which (under natural identification) is isomorphic to $\mathbb{R}^n$. Next, we define isometries.

Definition 4.2 (*isometry*) Suppose (M, g) and $(\tilde{M}, \tilde{g})$ are Riemannian manifolds with or without any boundary. An *isometry* $(M, g) \mapsto (\tilde{M}, \tilde{g})$ is a diffeomorphism (or an isomorphism over smooth manifolds), $\varphi : M \to \tilde{M}$ such that $\varphi^*\tilde{g} = g$.

Definition 4.2 is equivalent to the requirement that φ is a smooth bijection and each differential $d\varphi_p : T_pM \to T_{\varphi(p)}\tilde{M}$ be a linear isometry. In such a case, (M, g) and $(\tilde{M}, \tilde{g})$ are said to be isometric.

The most important tool that a Riemannian metric provides us is the ability to measure lengths of curves and distances between points. A curve in M always means a *parametrized curve*, that is, a continuous map $\gamma : I \mapsto M$, where $I \subseteq \mathbb{R}$ is some interval. A *curve segment* is a curve whose domain is a compact interval. Further, a *smooth curve* $\gamma : I \mapsto M$ has a well-defined velocity $\gamma'(t) \in T_{\gamma(t)}M$ for each $t \in I$. γ is a *regular curve* if $\gamma'(t) \neq 0$ for $t \in I$. We refer to a piecewise regular curve segment as an admissible curve. If $\gamma : [a, b] \mapsto M$ is an *admissible curve*, we define the length of γ to be,

$$L_g(\gamma) = \int_a^b |\gamma'(t)|_g dt.$$

Similarly, the *arc-length function of* γ is the function $s : [a, b] \mapsto \mathbb{R}$ defined by,

$$s(t) = L_g\left(\gamma|_{[a,t]}\right) = \int_a^t |\gamma'(t)|_g dt.$$

The above integrand is bounded and continuous everywhere on [a, b] except possibly at finitely many points where γ is not smooth, so the integral is well defined.

Proposition 4.1 (Properties of length) *Suppose* (M, g) *is a Riemannian manifold with or without any boundary and* $\gamma : I \mapsto M$ *is an admissible curve. The following properties hold:*
(a) Additivity of Length: If $a < c < b$, *then*

$$L_g(\gamma) = L_g\left(\gamma|_{[a,c]}\right) + L_g\left(\gamma|_{[c,b]}\right).$$

(b) Parameter independence: If $\tilde{\gamma}$ *is a reparametrization of* γ, *then* $L_g(\gamma) = L_g(\tilde{\gamma})$.

(c) Isometric invariance: If (M, g) *and* $(\tilde{M}, \tilde{g})$ *are Riemannian manifolds and* $\varphi : M \mapsto \tilde{M}$ *is a local isometry, then* $L_g(\gamma) = L_{\tilde{g}}(\varphi \circ \gamma)$.

Proposition 4.2 (Unit-speed reparametrization) *Suppose* (M, g) *is a Riemannian manifold with or without any boundary. Then, every regular curve in* M *has a unit-speed forward reparametrization.*

Proof Suppose $\gamma : I \mapsto M$ is a regular curve. Choose an arbitrary $t_0 \in I$, and define $s : I \mapsto \mathbb{R}$ by,

$$s(t) = \int_{t_0}^t |\gamma'(u)|_g du.$$

Since $s'(t) = |\gamma'(t)|_g > 0$, it follows that s is a strictly increasing local diffeomorphism, and thus maps I diffeomorphically onto an interval $I' \subseteq \mathbb{R}$. If we let $\varphi : s^{-1} : I' \mapsto I$, then $\tilde{\gamma} = \gamma \circ \varphi$ is a forward reparametrization of γ, and the chain rule gives,

$$|\tilde{\gamma}'(t)|_g = |\varphi'(s)\gamma'(s)|_g = \frac{1}{s'(\varphi(s))}|\gamma'(\varphi(s))|_g = 1.$$

Thus, $\tilde{\gamma}$ is a unit-speed reparametrization of γ. □

We are now in a position to introduce a very important concept from classical geometry into the Riemannian setting: distances between points. Suppose (M, g) is a connected Riemannian manifold with or without any boundary. For each pair of points $p, q \in M$, we define the *Riemannian distance from p to q*, denoted by $d_g(p, q)$ to be the infimum of the lengths of all admissible curves from p to q. The following proposition guarantees that $d_g(p, q)$ is a well-defined nonnegative real number for each $p, q \in M$.

Proposition 4.3 (Riemannian distance between points) *If M is a connected smooth manifold, then any two points of M can be joined by an admissible curve.*

Proof Let $p, q \in M$ be arbitrary. Since a connected manifold is path-connected, p and q can be connected by a continuous path $c : [a, b] \mapsto M$. By compactness, there is a partition of $[a, b]$ such that $c\,([a_{i-1}, a_i])$ is contained in a single smooth coordinate ball for each i. Then we may replace each such curve segment with a straight-line path in coordinates, yielding an admissible curve γ between the same points. □

A corollary follows from proposition 4.3.

Corollary 4.1 (Isometric invariance of the Riemannian Distance Function) *Suppose* (M, g) *and* $(\tilde{M}, \tilde{g})$ *are connected Riemannian manifolds and* $\varphi : M \mapsto \tilde{M}$ *is an isometry. Then* $d_{\tilde{g}}(\varphi(x), \varphi(y)) = d_g(x, y)$ *for all* $x, y \in M$.

4.2.1 Geodesics and Minimizing Curves

In this section, we investigate the relationships between geodesics, lengths and distances on a Riemannian manifold. Our primary goal is to show that all length-minimizing curves are geodesics and all geodesics are local length-minimizers. Let (M, g) be a Riemannian manifold. An admissible curve γ in M is said to be a *minimizing curve* if $L_g(\gamma) \leq L_g(\tilde{\gamma})$ for every admissible curve $\tilde{\gamma}$ with the same endpoints. When M is connected, it follows from the definition of Riemannian distance that γ is minimizing if and only if $L_g(\gamma)$ is equal to the distance between its endpoints. To show that all minimizing curves are geodesics, we will think of the length function L_g as a functional (real-valued functions whose domains are themselves sets of functions) on the set of all admissible curves in M with fixed starting and ending points. Our task is to find the minima of this functional. It suffices to note that if γ is a minimizing curve, and $\{\Gamma_s : s \in (-\epsilon, \epsilon)\}$ is a one-parameter family of admissible curves with the same endpoints such that $L_g(\Gamma_s)$ is a differentiable function of s and $\Gamma_0 = \gamma$, then by elementary calculus, the s-derivative of $L_g(\Gamma_s)$ must vanish at $s = 0$

because $L_g(\Gamma_s)$ attains a minimum there. To make these arguments more rigorous, we introduce more definitions.

Definition 4.3 (*variation of a curve*) If $\gamma : [a, b] \mapsto M$ is a given admissible curve, a *variation of* γ is an admissible family of curves $\Gamma : J \times [a, b] \mapsto M$ such that J is an open interval containing 0 and $\Gamma_0 = \gamma$. It is called a *proper variation* if, in addition, all of the main curves have the same starting and ending points: $\Gamma_s(a) = \gamma(a)$ and $\Gamma_s(b) = \gamma(b)$ for all $s \in J$.

Lemma 4.1 *If* γ *is an admissible curve and* V *is a piecewise smooth vector field along* γ*, then* V *is the variation field of some variation of* γ*. If* V *is proper, the variation can be taken to be proper as well.*

Proof Suppose γ and V satisfy the hypotheses, and set $\Gamma(s, t) = exp_{\gamma(t)}(sV(t))$ (where $\exp_p$ is the restriction of exponential function to the tangent space of the manifold, T_pM). By compactness of $[a, b]$, $\exists$ some positive ϵ such that Γ is defined on $(-\epsilon, \epsilon) \times [a, b]$. By composition, Γ is smooth on $(-\epsilon, \epsilon) \times [a_{i-1}, a_i]$ for each subinterval $[a_{i-1}, a_i]$ on which V is smooth, and continuous on its whole domain. Using the properties of the exponential map [13], the variation of Γ is V. Moreover, if $V(a) = 0$ and $V(b) = 0$, the definition gives $\Gamma(s, a) \equiv \gamma(a)$ and $\Gamma(s, b) \equiv \gamma(b)$, so Γ is proper. □

Lemma 4.2 (Symmetry Lemma) *Let* $\Gamma : J \times [a, b] \mapsto M$ *be an admissible family of curves in a Riemannian manifold. On every rectangle* $J \times [a_{i-1}, a_i]$ *where* Γ *is smooth, we have*

$$D_s \partial_t \Gamma = D_t \partial_s \Gamma.$$

Proof The derivatives are local, so we can compute in local coordinates of Γ around the point $\Gamma(s_0, t_0)$. Writing the components of Γ as $\Gamma(s, t) = \left(x^1(s, t), \ldots, x^n(s, t)\right)$, we have

$$\partial_t \Gamma = \frac{\partial x^k}{\partial t} \partial_k; \quad \partial_s \Gamma = \frac{\partial x^k}{\partial s} \partial_k.$$

Using the standard formula for contra variant derivatives along the curves [16], we obtain,

$$D_s \partial_t \Gamma = \left(\frac{\partial^2 x^k}{\partial s \partial t} + \frac{\partial x^i}{\partial t} \frac{\partial x^j}{\partial s} \Gamma_{ji}^k \right) \partial_k;$$
$$D_t \partial_s \Gamma = \left(\frac{\partial^2 x^k}{\partial t \partial s} + \frac{\partial x^i}{\partial s} \frac{\partial x^j}{\partial t} \Gamma_{ji}^k \right) \partial_k.$$

Reversing the roles of i and j in the second line above and using the symmetry condition $\Gamma_{ji}^k = \Gamma_{ij}^k$, we conclude that these two expressions are equal. □

Lemma 4.2 allows us to compute the expression for the derivative of the length functional along the variation of a curve. Typically, the derivative of a functional on a space of maps is called its *first variation*. One more result is required, as outlined below before we can show that all minimizing curves are geodesics.

Theorem 4.1 (First Variation Formula) *Let (M, g) be a Riemannian manifold. Suppose $\gamma : [a, b] \mapsto M$ is a unit-speed admissible curve $\Gamma : J \times [a, b] \mapsto M$ is a variation of γ, and V is its variation field. Then $L_g(\Gamma_s)$ is a smooth function of s, and*

$$\frac{d}{ds}|_{s=0} L_g(\Gamma_s) = -\int_a^b \langle V, D_t\gamma' \rangle dt - \sum_{i=1}^{k-1} \langle V(a_i), \Delta_i\gamma' \rangle + \langle V(b), \gamma'(b) \rangle - \langle V(a), \gamma'(a) \rangle, \quad (4.1)$$

where $(a_0, \ldots, a_k)$ is an admissible partition for V, and for each $i = 1, \ldots, k-1$, $\Delta_i\gamma' = \gamma'(a_i^+) - \gamma'(a_i^-)$ is the "jump" in the velocity vector field γ' at a_i. In particular, if Γ is a proper variation, then

$$\frac{d}{ds}|_{s=0} L_g(\Gamma_s) = -\int_a^b \langle V, D_t\gamma' \rangle dt - \sum_{i=1}^{k-1} \langle V(a_i), \Delta_i\gamma' \rangle. \quad (4.2)$$

Proof On every rectangle $J \times [a_{i-1}, a_i]$ where Γ is smooth, since the integrand in $L_g(\Gamma_s)$ is smooth and the domain of integration is compact, we can differentiate under the integral sign as many times as we wish. Because $L_g(\Gamma_s)$ is a finite sum of such integrals, it follows that it is a smooth function of s. For brevity, let us introduce the notations,

$$T(s, t) = \partial_t \Gamma(s, t), \quad S(s, t) = \partial_s \Gamma(s, t).$$

Differentiating on the interval $[a_{i-1}, a_i]$ gives,

$$\begin{aligned} \frac{d}{ds}|_{s=0} L_g(\Gamma_s|_{[a_{i-1}, a_i]}) &= \int_{a_{i-1}}^{a_i} \frac{\partial}{\partial s} \langle T, T \rangle^{1/2} dt, \\ &= \int_{a_{i-1}}^{a_i} \frac{1}{2} \langle T, T \rangle^{-1/2} 2 \langle D_s T, T \rangle dt, \\ &= \int_{a_{i-1}}^{a_i} \frac{1}{|T|} \langle D_t S, T \rangle dt, \end{aligned} \quad (4.3)$$

where Lemma 4.2 is used in the last line of Eq. (4.3). Setting $s = 0$ and noting that $S(0, t) = V(t)$ and $T(0, t) = \gamma'(t)$ (which has length 1), we find that,

$$
\begin{aligned}
\frac{d}{ds} L_g(\Gamma_s|_{[a_{i-1},a_i]}) &= \int_{a_{i-1}}^{a_i} \langle D_t V, \gamma' \rangle dt, \\
&= \int_{a_{i-1}}^{a_i} \left(\frac{d}{dt} \langle V, \gamma' \rangle - \langle V, D_t \gamma' \rangle \right) dt, \\
&= \langle V(a_i), \gamma'(a_i^-) \rangle - \langle V(a_{i-1}), \gamma'(a_{i-1}^+) \rangle - \int_{a_{i-1}}^{a_i} \langle V, D_t \gamma' \rangle dt.
\end{aligned}
\tag{4.4}
$$

where the second equality in Eq. (4.4) follows from the fundamental theorem of calculus. Finally, summing over i, we obtain (4.1). □

Because every admissible curve has a unit-speed parametrization and length is independent of parametrization, the requirement that the curve be of unit speed is not a real restriction. Next, comes the fundamental result characterizing the 'minimizing' nature of geodesics.

Theorem 4.2 *In a Riemannian manifold, every minimizing curve is a geodesic when it is given a unit-speed parametrization.*

Proof Let $\gamma : [a, b] \mapsto M$ is minimizing and of unit speed, and $(a_0, \ldots, a_k)$ is an admissible partition for γ. If Γ is any proper variation of γ, then $L_g(\Gamma_s)$ is a smooth function of s that achieves its minimum at $s = 0$, so it follows from standard calculus that $d\left(L_g(\Gamma_s)\right)/ds = 0$ when $s = 0$. Since every proper vector field along γ is the variation field of some proper variation, the right-hand side of equation (4.2) must vanish for every such V.

We first show that $D_t\gamma' = 0$ on each subinterval $[a_{i-1}, a_i]$, so γ is a "broken geodesic". Choose one such interval, and let $\varphi \in C^\infty(\mathbb{R})$ be a bump function such that $\varphi > 0$ on (a_{i-1}, a_i) and $\varphi = 0$ elsewhere. Then equation (4.2) with $V = \varphi D_t\gamma'$ becomes,

$$
-\int_{a_{i-1}}^{a_i} \varphi |D_t\gamma'|^2 dt = 0.
$$

Since the integrand is nonnegative and $\varphi > 0$ on (a_{i-1}, a_i), this shows that $D_t\gamma' = 0$ on each such subinterval.

Next, we show that $\Delta\gamma_i' = 0$ for each i between 0 and k, which is to say that γ has no corners. For each such i, we can use a bump function in a coordinate chart to construct a piecewise smooth vector field V along γ such that $V(a_i) = \Delta_i\gamma'$ and $V(a_i) = 0$ for $j \neq i$. Then Eq. (4.2) reduces to $-|\Delta\gamma'|^2 = 0$, so $\Delta_i\gamma' = 0$ for each i.

Finally, since the two one-sided velocity vectors of γ match up at each a_i, it follows from uniqueness of geodesics that $\gamma|_{[a_i,a_{i+1}]}$ is the continuation of the geodesic $\gamma|_{[a_{i-1},a_i]}$, and therefore γ is smooth. □

4.2.2 Completeness of Riemannian Manifolds

Suppose (M, g) is a connected Riemannian manifold Considering M as a metric space, it is time to address one of the most important questions one can ask about a metric space: Is it complete? In general, the answer is no: for example, if M is an open region in $\mathbb{R}^n$ with its Euclidean metric, then every sequence in M that converges in $\mathbb{R}^n$ to a point in ∂M is Cauchy, but not convergent in M.

A manifold is said to be *geodesically complete* if every maximal geodesic is defined for all $t \in \mathbb{R}$. For clarity, we will use the phrase *metrically complete* for a connected Riemannian manifold that is complete as a metric space with the Riemannian distance function, in the sense that every Cauchy sequence converges. The Hopf-Rinow Theorem, which we will state and prove below, shows that these two notions of completeness are equivalent for connected Riemannian manifolds. Before we prove it, let us establish a preliminary result, which will have other important consequences besides the Hopf-Rinow Theorem itself.

Lemma 4.3 *Suppose (M, g) is a connected Riemannian manifold, and there is a point $p \in M$ such that* $\exp_p$ *is defined on the whole tangent space T_pM.*
(a) Given any other $q \in M$, there is a minimizing geodesic segment from p to q.
(b) M is metrically complete.

Proof Let $q \in M$ be arbitrary. If $\gamma : [a, b] \mapsto M$ is a geodesic segment starting at p, let us say that γ *aims at* q if γ is minimizing and,

$$d_g(p, q) = d_g(p, \gamma(b)) + d_g(\gamma(b), q). \tag{4.5}$$

To prove (a), it suffices to show that there is a geodesic segment $\gamma : [a. b] \mapsto M$ that begins at p, aims at q, and has length equal to $d_g(p, q)$, for then the fact that γ is minimizing means that $d_g(p, \gamma(b)) = L_g(\gamma) = d_g(p, q)$, and Eq. (4.5) becomes,

$$d_g(p, q) = d_g(p, q) + d_g(\gamma(b). q),$$

which implies $\gamma(b) = q$. Since γ is a segment from p to q of length $d_g(p, q)$, it is the desired minimizing geodesic segment.

Choose $\epsilon > 0$ such that there is a closed geodesic sphere, $B_\epsilon(p)$, around p that does not contain q. Since the distance function on a metric space is continuous, there is a point x in the geodesic sphere $S_\epsilon(p)$ where $d_g(x, q)$ attains its minimum on the compact set $S_\epsilon(p)$. Let γ be the maximal unit-speed geodesic whose restriction to $[0, \epsilon]$ is the radial geodesic segment from p to x; by assumption, γ is defined for all $t \in \mathbb{R}$.

We begin by presenting that $\gamma|_{[0,\epsilon]}$ aims at q. Since every closed geodesic ball is contained in a larger open one, we need to only show that Eq. (4.5) holds with $b = \epsilon$, or

$$d_g(p, q) = d_g(p, x) + d_g(x, q). \tag{4.6}$$

To this end, let $\sigma : [a_0, b_0] \mapsto M$ be any admissible curve from p to q. Let t_0 be the first time σ hits $S_\epsilon(p)$, and let σ_1 and σ_2 denote the restrictions of σ_1 to $[a_0, t_0]$ and $[t_0, b_0]$, respectively. Since every point in $S_\epsilon(p)$ is at a distance ϵ from p, we have $L_g(\sigma_1) \geq d_g(p, \sigma(t_0)) = d_g(p, x)$; and by our choice of x we have $L_g(\sigma_2) \geq d_g(\sigma(t_0), q) \geq d_g(x, q)$. Putting these two inequalities together yields,

$$L_g(\sigma) = L_g(\sigma_1) + L_g(\sigma_2) \geq d_g(p, x) + d_g(x, q).$$

Taking the infimum over all such σ, we find that $d_g(p, q) \geq d_g(p, x) + d_g(x, q)$. The opposite inequality is just the triangle inequality, so Eq. (4.6) holds. Now let $T = d_g(p, q)$ and

$$\mathcal{A} = \{b \in [0, T] : \gamma|_{[0,b]} \text{aims at} q\}.$$

We have just shown that $\epsilon \in \mathcal{A}$. Let $A = \sup\mathcal{A} \geq \epsilon$. By continuity of the distance function, it is easy to see that $\mathcal{A}$ is closed in $[0, T]$, and therefore $A \in \mathcal{A}$. If $A = T$, then $\gamma|_{[0,T]}$ is a geodesic of length $T = d_g(p, q)$ that aims at q, and by the remark above we are done. So we assume $A < T$ and derive a contradiction.

Let $y = \gamma(A)$, and choose $\delta > 0$ such that there is a closed geodesic sphere, $B_\delta(y)$, around y, small enough that it does not contain q. The fact that $A \in \mathcal{A}$ means that

$$d_g(y, q) = d_g(p, q) - d_g(p, y) = T - A.$$

Let $z \in S_\delta(y)$ be a point where $d_g(z, q)$ attains its minimum, and let $\tau : [0, \delta] \mapsto M$ be the unit-speed radial geodesic from y to z. Using exactly the same argument as earlier, τ aims at q, so,

$$d_g(z, q) = d_g(y, q) - d_g(y, z) = (T - A) - \delta. \tag{4.7}$$

Using triangle inequality and Eq. (4.7),

$$d_g(p, z) \geq d_g(p, q) - d_g(z, q) = T - (T - A - \delta) = A + \delta.$$

Therefore, the admissible curve consisting of $\gamma|_{[0,A]}$ (of length A) followed by τ (of length δ) is a minimizing curve from p to z. This means that it has no corners, so z must lie on γ, and in fact, $z = \gamma(A + \delta)$. But, Eq. (4.7) says that,

$$d_g(p, q) = T = (A + \delta) + d_g(z, q) = d_g(p, z) + d_g(z, q),$$

so $\gamma|_{[0,A+\delta]}$ aims at q and $A + \delta \in \mathcal{A}$, which is a contradiction. This completes the proof of (a).

To prove (b), we need to show that every Cauchy sequence in M converges. Let (q_i) be a Cauchy sequence in M. For each i, let $\gamma_i(t) = \exp_p(tv_i)$ be a unit-speed minimizing geodesic from p to q_i, and let $d_i = d_g(p, q_i)$, so that $q_i = \exp_p(d_i v_i)$. The sequences (d_i) is bounded in $\mathbb{R}$ (because Cauchy sequences in a metric space are bounded), and the sequence (v_i) consists of unit vectors in T_pM, so the sequence of vectors (d_i, v_i) in T_pM is bounded. Therefore a subsequence (d_{i_k}, v_{i_k}) converges to some $v \in T_pM$. By continuity of the exponential map, $q_{i_k} = \exp_p(d_{i_k} v_{i_k}) \mapsto \exp_p v$, and since the original sequence (q_i) is Cauchy, it converges to the same limit. □

The next theorem is the main result of this section.

Theorem 4.3 (Hopf-Rinow) *A connected Riemannian manifold is metrically complete if and only if it is geodesically complete.*

Proof Let (M, g) be a connected Riemannian manifold. Suppose first that M is geodesically complete. Then, in particular, it satisfies the hypothesis of Lemma 4.3, so it is metrically complete.

Conversely, suppose M is metrically complete, and assume for the sake of contradiction that it is not geodesically complete. Then there is some unit-speed geodesic $\gamma : [0, b) \mapsto M$ that has no extension to a geodesic on any interval $[0, b')$ for $b' > b$. Let (t_i) be any increasing sequence in $[0, b)$ that approaches b, and set $q_i = \gamma_i(t)$. Since γ is parametrized by arc length, the length of $\gamma|_{[t_i, t_j]}$ is exactly $|t_j - t_i|$, so $d_g(q_i, q_j) \leq |t_j - t_i|$ and (q_i) is a Cauchy sequence in M. By completeness, (q_i) converges to some point $q \in M$.

Let W be a uniformly $\delta-$normal neighborhood of q for some $\delta > 0$. Choose j large enough that $t_j > b - \delta$ and $q_j \in W$. The fact that $B_\delta(q_j)$ is a geodesic ball means that every unit-speed geodesic starting at q_j exists at least for $t \in [0, \delta)$. In particular, this is true of the geodesic σ with $\sigma(0) = q_j$ and $\sigma'(0) = \gamma'(t_j)$. Define $\tilde{\gamma} : [0, t_j + \delta) \mapsto M$ by

$$\tilde{\gamma}(t) = \begin{cases} \gamma(t), & t \in [0, b), \\ \sigma(t - t_j), & t \in (t_j - \delta, t_j + \delta). \end{cases}$$

Note that both expressions on the right-hand side are geodesics, and they have the same position and velocity when $t = t_j$. Therefore, by the uniqueness of geodesics, the two definitions agree where they overlap. Since $t_j + \delta > b$, $\tilde{\gamma}$ is an extension of γ past b, which is a contradiction. □

We conclude this section by stating three important corollaries, whose proofs are easy applications of Lemma 4.3 and the Hopf-Rinow Theorem 4.3.

Corollary 4.2 *If M is a connected Riemannian manifold and there exists a point $p \in M$ such that the restricted exponential map $\exp_p$ is defined on all of T_pM, then M is complete.*

Corollary 4.3 *If M is a complete, connected Riemannian manifold, then any two points in M can be joined by a minimizing geodesic segment.*

Corollary 4.4 *If M is a compact Riemannian manifold, then every maximal geodesic in M is defined for all time.*

4.3 Mathematical Model: Streamfunction-Vorticity Formulation

We briefly digress to rederive the mathematical model for incompressible, subdiffusive, planar, viscoelastic channel flows in the streamfunction-vorticity formulation. The advantage of investigating the equations in this form is that the original system of partial differential equations in velocity-pressure formulation (containing three equations in two dimensions) is converted into the reduced set of two equations, which is free of the variable, pressure. The continuity equation is identically satisfied in streamfunction-vorticity formulation.

The scalar streamfunction, ψ, is defined for incompressible and divergence-free flows in two dimensions. The difference between the stream function values at any two points gives the volumetric flow rate (or volumetric flux) through a line connecting the two points. Hence, given the velocity field $\mathbf{v} = (u, v)$, then,

$$u = \frac{\partial \psi}{\partial y}, \qquad v = -\frac{\partial \psi}{\partial x}. \tag{4.8}$$

Vorticity, Ω, is a measure of the local rotation of a fluid element and is defined as the curl of the fluid velocity vector. Vorticity has only one non-zero component for two-dimensional flows, and this component points out of the plane of the flow. Mathematically, vorticity can be expressed as,

$$\Omega = (\nabla \times \mathbf{v}) \cdot \hat{k} = \frac{\partial v}{\partial x} - \frac{\partial u}{\partial y}. \tag{4.9}$$

Recall the component-wise momentum equations in the velocity-pressure formulation (Eq. (3.34)),

$$Re\left[\frac{\partial^\alpha u}{\partial t^\alpha} + u\frac{\partial u}{\partial x} + v\frac{\partial u}{\partial y}\right] = \nu\left[\frac{\partial^2 u}{\partial x^2} + \frac{\partial^2 u}{\partial y^2}\right] + \frac{(1-\nu)}{We}\left[\frac{\partial C_{11}}{\partial x} + \frac{\partial C_{12}}{\partial y}\right], \tag{4.10a}$$

$$Re\left[\frac{\partial^\alpha v}{\partial t^\alpha} + u\frac{\partial v}{\partial x} + v\frac{\partial v}{\partial y}\right] = \nu\left[\frac{\partial^2 v}{\partial x^2} + \frac{\partial^2 v}{\partial y^2}\right] + \frac{(1-\nu)}{We}\left[\frac{\partial C_{12}}{\partial x} + \frac{\partial C_{22}}{\partial y}\right]. \tag{4.10b}$$

Differentiating equation (4.10b) partially with respect to x, and differentiating equation (4.10a) partially with respect to y, and subtracting (assume sufficient smoothness to permit changing the order of differentiation), we note that,

$$Re\left[\frac{\partial^\alpha}{\partial t^\alpha}\left(\frac{\partial v}{\partial x}-\frac{\partial u}{\partial y}\right)+\left(u\frac{\partial}{\partial x}+v\frac{\partial}{\partial y}\right)\left(\frac{\partial v}{\partial x}-\frac{\partial u}{\partial y}\right)\right]$$
$$=\nu\left[\frac{\partial^2}{\partial x^2}\left(\frac{\partial v}{\partial x}-\frac{\partial u}{\partial y}\right)+\frac{\partial^2}{\partial y^2}\left(\frac{\partial v}{\partial x}-\frac{\partial u}{\partial y}\right)\right]$$
$$+\frac{(1-\nu)}{We}\left[\frac{\partial^2\mathcal{C}_{12}}{\partial x^2}-\frac{\partial^2\mathcal{C}_{12}}{\partial y^2}+\frac{\partial^2}{\partial x\partial y}(\mathcal{C}_{22}-\mathcal{C}_{11})\right]. \tag{4.11}$$

From the definitions of streamfunction (Eq. (4.8)) and vorticity (Eq. (4.9)), we arrive at the Poisson equation,

$$\Omega=\frac{\partial v}{\partial x}-\frac{\partial u}{\partial y}=-\frac{\partial^2\psi}{\partial x^2}-\frac{\partial^2\psi}{\partial y^2}=-\nabla^2\psi. \tag{4.12}$$

Similarly, substituting $\mathbf{A}=\frac{\mathcal{C}-\mathbf{I}}{We}$ in Eq. (3.32), we find that,

$$\frac{\partial^\alpha}{\partial t^\alpha}\left(\frac{\mathcal{C}-\mathbf{I}}{We}\right)+\mathbf{v}\cdot\nabla\left(\frac{\mathcal{C}-\mathbf{I}}{We}\right)-(\nabla\mathbf{v})^T\left(\frac{\mathcal{C}-\mathbf{I}}{We}\right)-\left(\frac{\mathcal{C}-\mathbf{I}}{We}\right)\nabla\mathbf{v}=\frac{\mathbf{D}-\left(\frac{\mathcal{C}-\mathbf{I}}{We}\right)}{We}, \tag{4.13}$$

which implies

$$\frac{\partial^\alpha\mathcal{C}}{\partial t^\alpha}+\mathbf{v}\cdot\nabla\mathcal{C}-(\nabla\mathbf{v})^T\mathcal{C}+(\nabla\mathbf{v})^T-\mathcal{C}(\nabla\mathbf{v})+(\nabla\mathbf{v})=(\nabla\mathbf{v})+(\nabla\mathbf{v})^T-\left(\frac{\mathcal{C}-\mathbf{I}}{We}\right). \tag{4.14}$$

Using Eqs. (4.11), (4.12) and (4.14), we obtain the model in streamfunction-vorticity form, as

$$Re\left[\frac{\partial^\alpha\Omega}{\partial t^\alpha}+\mathbf{v}\cdot\nabla\Omega\right]=\nu\nabla^2\Omega+\frac{(1-\nu)}{We}\nabla\times\nabla\cdot\mathcal{C}, \tag{4.15a}$$
$$\nabla^2\psi=-\Omega, \tag{4.15b}$$
$$\frac{\partial^\alpha\mathcal{C}}{\partial t^\alpha}+\mathbf{v}\cdot\nabla\mathcal{C}-(\nabla\mathbf{v})^T\mathcal{C}-\mathcal{C}\nabla\mathbf{v}=\frac{\mathbf{I}-\mathcal{C}}{We}, \tag{4.15c}$$

where $\frac{\partial^\alpha f(\mathbf{x},t)}{\partial t^\alpha}$ denotes the Caputo fractional derivative of order α [14] with respect to t defined by,

$$\frac{\partial^\alpha f(\mathbf{x},t)}{\partial t^\alpha}=\frac{1}{\Gamma(1-\alpha)}\int_0^t\frac{dt'}{(t-t')^\alpha}\frac{\partial f(\mathbf{x},t')}{\partial t'},\quad 0<\alpha<1, \tag{4.16}$$

and the operators $\nabla(\cdot)$ and $\nabla^2(\cdot)$ in Eq. (4.15a), are (integer order) gradient and Laplacian operators in $\mathbb{R}^2$. Note that Eq. (4.15c) represents the fractional version of the regular Oldroyd-B model for viscoelastic fluids [17] in terms of the conformation tensor $\mathcal{C}$ obtained from the equation for elasticity tensor $\mathbf{A}$ using the relation $\mathcal{C} = We\mathbf{A} + I$ (3.35).

From the perspective of continuum mechanics, $\mathcal{C}$, is the Finger tensor associated with polymer deformation [10], such that

$$\mathcal{C} = \boldsymbol{F}\boldsymbol{F}^T, \tag{4.17}$$

where $\boldsymbol{F} = \boldsymbol{F}(t)$ is the instantaneous deformation gradient tensor. If the spatial coordinates in the micro-structure are given by $\boldsymbol{r} = \boldsymbol{r}(\boldsymbol{r}_0, t)$ where $\boldsymbol{r}_0$ are the coordinates at equilibrium, then $\boldsymbol{F} = \partial\boldsymbol{r}/\partial\boldsymbol{r}_0$. In other words, a vector $d\boldsymbol{r}_0$ deforms to $d\boldsymbol{r} = \boldsymbol{F}d\boldsymbol{r}_0$ under the deformation, $\boldsymbol{F}$.

4.4 Dynamics of Structure Tensor

Now, we resume our efforts to develop an appropriate measure to visualize macrostructures. The caveats in the conformation tensor outlined in Sect. 4.1 (namely, the loss of positive definiteness in the arithmetic compilation of fluctuations and unequal measure for equally probable states representing contraction and expansion) enforce us to adopt a different framework to capture polymer deformation in sub-diffusive flows. We begin by denoting the general linear group of degree n, which is the set of all $n \times n$ invertible matrices, as $\boldsymbol{GL}_n$.

Definition 4.4 Define the structure-preserving group action of $\boldsymbol{GL}_n$ on a set $\mathbf{V}_n \subseteq \mathbb{R}^{n\times n}$ as,

$$[\boldsymbol{B}]_{\boldsymbol{A}} \equiv \boldsymbol{A}\boldsymbol{B}\boldsymbol{A}^T,$$

where $\boldsymbol{A} \in \boldsymbol{GL}_n$ and $\boldsymbol{B} \in \mathbf{V}_n$.

Using definition (4.4), we find that Eq. (4.17) reduces to

$$\mathcal{C} = [\boldsymbol{I}]_{\boldsymbol{F}}. \tag{4.18}$$

Let $\overline{\mathcal{C}}$ be the mean conformation tensor (or the conformation tensor associated with the flow at equilibrium, then we assume that $\mathcal{C}$ is similar to $\overline{\mathcal{C}}$ under the group action (4.4) for any rotation matrix $\boldsymbol{S} \in \boldsymbol{SO}_n \subseteq \boldsymbol{GL}_n$, where $\boldsymbol{SO}_n$ represents the $n \times n$ special orthogonal group of rotation matrices.

Similarly, define $\overline{\boldsymbol{F}} \in \boldsymbol{GL}_n$ as the deformation gradient tensor associated with the mean configuration such that,

$$\overline{\mathcal{C}} = \overline{\boldsymbol{F}}\overline{\boldsymbol{F}}^T. \tag{4.19}$$

We remark that $\overline{\boldsymbol{F}}$ is non-unique since it can be represented as

$$\overline{\boldsymbol{F}} = \overline{\mathcal{C}}^{1/2}\boldsymbol{S}, \tag{4.20}$$

for any $S \in SO_3$. $\overline{\mathcal{C}}^{1/2}$ is the unique matrix square-root, found exclusively in terms of $\overline{\mathcal{C}}$ and its invariants (i. e., its trace and determinants) using an application of the representation theorem [11]. Since all we require is that the det $\overline{F} > 0$ (in order to maintain the positive definiteness of $\overline{\mathcal{C}}$), we choose $S = I$ in Eq. (4.20).

Given $\overline{F}$ satisfying equation (4.20), we can decompose the instantaneous deformation gradient tensor F and $\overline{F}$ by considering successive transformations on the vector dr_0 as,

$$dr = Fdr_0 = \overline{F}\mathcal{L}dr_0, \tag{4.21}$$

where $\mathcal{L} = \overline{F}^{-1}F$ is the tensor describing fluctuations away from the mean configuration, denoted as fluctuating deformation gradient tensor. Alternatively, substituting $F = \overline{F}\mathcal{L}$ in Eq. (4.17) and utilizing definition (4.4), we arrive at the following definition,

Definition 4.5 (*Structure Tensor*) Define $\mathcal{G}$ such that

$$\mathcal{C} = \overline{F}\mathcal{G}\overline{F}^T = [\mathcal{G}]_{\overline{F}},$$

where $\mathcal{G} = \mathcal{L}\mathcal{L}^T$.

The fluctuating conformation tensor, $\mathcal{C}'(= \mathcal{C} - \overline{\mathcal{C}})$ is related to the Structure Tensor as follows,

$$\mathcal{C}' = [\mathcal{G} - I]_{\overline{F}}. \tag{4.22}$$

Using definition (4.5) in Eq. (4.15c) and pre multiplying (post multiplying) the resultant equation by $\overline{F}^{-1}$ $(\overline{F}^{-T})$ and noting that $\mathcal{G}$ is a symmetric tensor, we arrive at the following equation governing the dynamics of Structure Tensor,

$$\frac{\partial^\alpha \mathcal{G}}{\partial t^\alpha} + \mathbf{v} \cdot \nabla\mathcal{G} = \mathcal{G}\mathfrak{F}(\mathbf{v}) + \mathfrak{F}(\mathbf{v})^T\mathcal{G} - \mathcal{M}, \tag{4.23}$$

which replaces Eq. (4.15c) in the viscoelastic sub-diffusive model (4.15). The functions, $\mathfrak{F}(\mathbf{v})=\overline{F}^T\nabla\mathbf{v}\overline{F}^{-T} - \left(\overline{F}^{-1}(\mathbf{v}\cdot\nabla)\,\overline{F}\right)^T$, and $\mathcal{M}=\frac{1}{W_2}\left(\mathcal{G}-\left(\overline{F}^T\overline{F}\right)^{-1}\right)$.

4.5 Scalar Invariants Via a Riemannian Geodesic

The tensorial nature of $\mathcal{G}$ renders the quantification of the fluctuating conformation tensor, a difficult task. By utilizing tr $\mathcal{G}$, Berris [10] made an initial attempt to characterize polymer deformation in the Oldroyd-B model by defining a 'elastic potential energy'. Elastic energy was an insufficient descriptor in characterizing polymer deformation due to (a) its dependence on the choice of the particular constitutive model, and (b) elastic energy was found to be the same for a family of conformation tensors with identical trace but variable determinant. We instead evolve an approach to characterize deformation using the inherent structure of the tensor $\mathcal{G}$.

Any scalar characterization of $\mathcal{G}$ can be naively developed as a function of its three principle invariants, i.e., trace, dyadic product of eigenvalues and determinant. However, even for simple cases (such as the isotropic case), the invariants are bounded between 0 and 1 (1 and ∞) for compression with respect to $\overline{\mathcal{C}}$ (expansion with respect to $\overline{\mathcal{C}}$). This asymmetric characterization is undesirable. Further, the statistical moments of the invariants vary over several orders of magnitude, rendering these moments as uninformative predictors of polymer stretching. The above-mentioned problems arise because the set of $n \times n$ positive definite matrices (denoted with $\boldsymbol{PS}_n$ in subsequent discussion) do not form a vector space, and thus, the Euclidean notion of translation and distances are irrelevant. Instead, we exploit the Riemannian structure of $\boldsymbol{PS}_n$ to formulate alternative scalar measures of $\mathcal{G}$.

$\boldsymbol{PS}_n$ is a Hilbert space where we can define an inner product given by $\langle \boldsymbol{A}, \boldsymbol{B} \rangle_X = \mathrm{tr}\left(\boldsymbol{X}^{-1}\boldsymbol{A}^T\boldsymbol{X}^{-1}\boldsymbol{B}\right)$, and a corresponding induced norm $\|\boldsymbol{A}\|_X = \left(\mathrm{tr}\left(\boldsymbol{X}^{-1}\boldsymbol{A}^T\boldsymbol{X}^{-1}\boldsymbol{A}\right)\right)^{1/2}$, where $\boldsymbol{X} \in \boldsymbol{PS}_n$. Furthermore, since $\boldsymbol{PS}_n$ is an open subset of the space of $n \times n$ real-valued matrices, it is a differentiable manifold. Using a simple argument, it can be shown that the tangent space at every point in $\boldsymbol{PS}_n$ is the space of symmetric matrices. However, $\boldsymbol{PS}_n$ can be shown to be a Riemannian manifold with a geodesic, which is obtained via the same inner product (defined above) on the tangent space at every point. The next set of results forms the requisite machinery to formulate this geodesic, which will be needed to define the scalar invariants of $\mathcal{G}$.

Consider a parametrized curve on $\boldsymbol{PS}_n$ connecting points $\boldsymbol{X}, \boldsymbol{Y} \in \boldsymbol{PS}_n$. That is $P : [0, 1] \rightarrow \boldsymbol{PS}_n$ with $P(0) = \boldsymbol{X}$ and $P(1) = \boldsymbol{Y}$. The distance, in the sense of the Riemannian metric, traversed on the manifold along the curve $P = P(r)$ is given by:

$$\ell_P(r) = \int_0^r \left\| \frac{dP(r')}{dr'} \right\|_{P(r')} dr'.$$

ℓ_P is invariant under affine transformation, as shown in the next lemma.

Lemma 4.4 (Affine invariance [11]) *For every positive definite matrix* $\boldsymbol{A}$ *and differentiable path* P *on the Riemannian manifold of positive definite matrices, we have:*

$$\ell_P = \ell_{[P]_A},$$

where $[\cdot]_A$ *denotes an action under* $\boldsymbol{A}$ *of the form* $\boldsymbol{A}^T\boldsymbol{P}\boldsymbol{A}$ *(see Definition 4.4 for details).*

Proof We use the definition of the norm $\|\cdot\|$ as stated above and the commutativity of the trace of the matrix product, to arrive at,

$$\begin{aligned}\|dP_{[A]}\|_{P_{[A]}} &= \left\|\left(A^T P(r) A\right)^{-1/2}\left(A^T P(r) A\right)'\left(A^T P(r) A\right)^{-1/2}\right\|_I \\ &= \left(\operatorname{tr}\left(\left(A^T P(r) A\right)^{-1}\left(A^T P(r) A\right)'\left(A^T P(r) A\right)^{-1}\left(A^T P(r) A\right)'\right)\right)^{1/2}, \\ &= \left(\operatorname{tr} A^{-1}(P)^{-1}(r) P'(r)(P)^{-1}(r) P'(r) A\right)^{1/2}, \\ &= \left(\operatorname{tr}(P)^{-1}(r) P'(r)(P)^{-1}(r) P'(r)\right)^{1/2}, \\ &= \left\|(P)^{-1/2}(r) P'(r)(P)^{-1/2}(r)\right\|_I = \|dP\|_P,\end{aligned}$$

where $'$ denotes the derivative with respect to the independent variable. The proof is completed by integrating both sides of this equality over r to obtain that,

$$\ell_P = \int_0^r \left\|\frac{dP(r')}{dr'}\right\|_{P(r')} dr' = \int_0^r \left\|\frac{dP_{[A]}(r')}{dr'}\right\|_{P_{[A]}(r')} dr' = \ell_{[P]_A}.$$

□

In the derivation above, we have used the cyclical property of the trace and the fact that an infinitesimal distance away from the point X on the manifold is given by $\|dX\|_X = \|X^{-1/2} dX X^{-1/2}\|_I = \left(\operatorname{tr}\left(X^{-1} dX\right)^2\right)^{1/2}$. Using Lemma 4.4, we can define $d(X, Y)$, (or the geodesic distance between X and Y) as the infimum of $\ell_P(1)$ over all possible curves P connecting X and Y as follows,

Definition 4.6

$$d(X, Y) = \inf_P \{\ell_P(1) \mid P(r) \in PS_n,\ P(0) = X,\ P(1) = Y\}.$$

A corollary of Lemma 4.4 is that $d(X, Y) = d([X]_A, [Y]_A)$. Note that the Hopf-Rinow Theorem guarantees the existence and uniqueness of such a geodesic. The next set of three results allows the construction of this geodesic.

Theorem 4.4 (Exponential metric increasing property [13]) *For any two real symmetric matrices X and Y, we have that:*

$$\|X - Y\|_I \le d(e^X, e^Y), \tag{4.24}$$

where e^X, e^Y are positive definite matrices.

Proof In order to demonstrate the proof, we wish to show the following inequality,

$$\|X - Y\|_I \le \left\|e^{-A/2}\left(De^A(X - Y)\right)e^{-A/2}\right\|_I, \tag{4.25}$$

where A is any real symmetric matrix, e^{-A} is the exponential map evaluated at the point A in the Riemannian manifold of symmetric matrices, and $De^A(X - Y)$ is the derivative of the exponential map at the point $\boldsymbol{A}$ evaluated on the matrix $\boldsymbol{X}-\boldsymbol{Y}$ and defined as follows,

Definition 4.7

$$De^{\boldsymbol{H}}(\boldsymbol{K}) := \lim_{t\to 0}\frac{e^{\boldsymbol{H}+t\boldsymbol{K}}-e^{\boldsymbol{H}}}{t},$$

for any matrices $\boldsymbol{H}$ and $\boldsymbol{K}$. The inequality (4.24) follows from the inequality (4.25) as follows. Let $\boldsymbol{H}(t)$ be any path joining symmetric matrices $\boldsymbol{X}$ and $\boldsymbol{Y}$, then $e^{\boldsymbol{H}}$ is the path joining $e^{\boldsymbol{X}}$ and $e^{\boldsymbol{Y}}$. Let $\chi = e^{\boldsymbol{H}}$ then $\chi' = De^{\boldsymbol{H}}(\boldsymbol{H}'(t))$. The length of this path is given by,

$$\begin{aligned}\ell_\chi = \int_0^1 \|d\chi\|_\chi &= \int_0^1 \left\|\chi^{-1/2}d\chi\chi^{-1/2}\right\|_{\boldsymbol{I}} dr\\ &= \int_0^1 \left\|e^{-\boldsymbol{H}/2}De^{\boldsymbol{H}}\boldsymbol{H}'(r)e^{-\boldsymbol{H}/2}\right\|_{\boldsymbol{I}} dr\\ &\geq \int_0^1 \left\|\boldsymbol{H}'(r)\right\|_{\boldsymbol{I}} dr\\ &\geq \|\boldsymbol{X}-\boldsymbol{Y}\|_{\boldsymbol{I}},\end{aligned} \tag{4.26}$$

where the last inequality appears because $\|\boldsymbol{X}-\boldsymbol{Y}\|_{\boldsymbol{I}}$ is the length of this path in the Euclidean space of symmetric matrices. But $d(e^{\boldsymbol{X}}, e^{\boldsymbol{Y}}) = \inf \ell_\chi \geq \|\boldsymbol{X}-\boldsymbol{Y}\|_{\boldsymbol{I}}$. (from Definition 4.6).

Thus, in order to prove inequality (4.25) we shall equivalently show for a real symmetric matrix, $\boldsymbol{B} = (\boldsymbol{X}-\boldsymbol{Y})$ that,

$$\|\boldsymbol{B}\|_{\boldsymbol{I}} \leq \left\|e^{-\boldsymbol{A}/2}\left(De^{\boldsymbol{A}}(\boldsymbol{B})\right)e^{-\boldsymbol{A}/2}\right\|_{\boldsymbol{I}}. \tag{4.27}$$

Choosing an orthonormal basis in which $\boldsymbol{A}$ =diag$(\lambda_1,\ldots,\lambda_n)$ and deploying the spectral decomposition formula (abridged from equation (2.40) in [11]), we have that,

$$De^{\boldsymbol{A}}(\boldsymbol{B}) = \frac{e^{\lambda_i}-e^{\lambda_j}}{\lambda_i-\lambda_j}b_{ij},$$

where b_{ij} is the i, j entry of the matrix, $\boldsymbol{B}$. Similarly, the i, j entry of the matrix, $e^{-\boldsymbol{A}/2}\left(De^{\boldsymbol{A}}(\boldsymbol{B})\right)e^{-\boldsymbol{A}/2}$, is,

$$\frac{\sinh(\lambda_i-\lambda_j)/2}{(\lambda_i-\lambda_j)/2}b_{ij}.$$

Since $(\sinh x)/x \geq 1$ for all real x, the inequality (4.27) follows. □

We note that the equality in equation (4.24) is achieved when $\boldsymbol{X}$ and $\boldsymbol{Y}$ commute, and in this special case, we can parametrize the geodesic, as outlined in the next result.

Proposition 4.4 *Let $X = e^{\mathcal{X}}$ and $Y = e^{\mathcal{Y}}$ be positive definite matrices such that $XY = YX$. Then, the exponential function maps the line segment [11],*

$$(1-r)\mathcal{X} + r\mathcal{Y}, \quad 0 \le r \le 1,$$

in the Euclidean space of symmetric matrices to the geodesic between X and Y on the Riemannian manifold of positive definite matrices and

$$d(X, Y) = \|\mathcal{X} - \mathcal{Y}\|_{I}. \tag{4.28}$$

Proof It is enough to show that the path given by,

$$\gamma(r) = e^{((1-r)\mathcal{X} + r\mathcal{Y})}, \quad 0 \le r \le 1, \tag{4.29}$$

is the unique path of shortest length joining X and Y in the space of symmetric matrices. Adopting the following parametrization of the path, $\gamma(r) = X^{1-r}Y^{r}$ as well as the commutativity of X and Y, we have,

$$\gamma'(r) = (\mathcal{Y} - \mathcal{X})\,\gamma(r). \tag{4.30}$$

The length of this path is given by (see Eq. (4.26) above),

$$\ell_{\gamma} = \int_0^1 \|\gamma^{-1/2} d\gamma \gamma^{-1/2}\|_{I}\, dr = \|\mathcal{X} - \mathcal{Y}\|_{I}. \tag{4.31}$$

Theorem 4.4 says that γ is the shortest path. All that remains to show is that the path γ under consideration is unique. Suppose $\tilde{\gamma}$ is another path that joins X and Y and has the same length as that of γ. Then $\log \tilde{\gamma}$ is a path joining $\mathcal{X}$ and $\mathcal{Y}$, and by Eq. (4.26) this path has minimum length $\|\mathcal{X} - \mathcal{Y}\|_{I}$. However, in an Euclidean space, the straight line segment is the unique shortest path between two points. Hence, the result follows. □

Finally, using the affine-invariance property (4.4) of the Riemannian metric and noting that I commutes with every element of PS_n, we arrive at the following general result.

Theorem 4.5 *Let X and Y be positive definite matrices. There exists a unique geodesic $X\#_r Y$ on the Riemannian manifold of positive definite matrices that joins X and Y with the following parametrization [11],*

$$X\#_r Y = X^{1/2}\left(X^{-1/2} Y X^{-1/2}\right)^{r} X^{1/2}, \tag{4.32}$$

which is natural in the sense that,

$$d(X, X\#_r Y) = r\, d(X, Y), \tag{4.33}$$

for each $r \in [0, \ 1]$. *Furthermore, we have,*

$$d(\boldsymbol{X}, \boldsymbol{Y}) = \left\| \log \left(\boldsymbol{X}^{-1/2} \boldsymbol{Y} \boldsymbol{X}^{1/2} \right) \right\|_{\boldsymbol{I}} = \left[\sum_{i=1}^{3} \left(\log \sigma_i \left(\boldsymbol{X}^{-1} \boldsymbol{Y} \right) \right)^2 \right]^{1/2}, \tag{4.34}$$

where σ_i *are the eigenvalues of the matrix* $\boldsymbol{X}^{-1}\boldsymbol{Y}$.

Proof Clearly, the matrices $\boldsymbol{I}$ and $\boldsymbol{X}^{-1/2}\boldsymbol{Y}\boldsymbol{X}^{-1/2}$ commute. Hence, the geodesic joining these two points is naturally parameterized as:

$$P_0(r) = \left(\boldsymbol{X}^{-1/2} \boldsymbol{Y} \boldsymbol{X}^{-1/2} \right)^r. \tag{4.35}$$

Applying the isometry $\boldsymbol{X}^{1/2}\boldsymbol{Y}\boldsymbol{X}^{1/2}$, we obtain the path,

$$P(r) = \boldsymbol{X}^{1/2} \left(P_0(r) \right) \boldsymbol{X}^{1/2} = \boldsymbol{X}^{1/2} \left(\boldsymbol{X}^{-1/2} \boldsymbol{Y} \boldsymbol{X}^{-1/2} \right)^r \boldsymbol{X}^{1/2}, \tag{4.36}$$

joining the points $\boldsymbol{X}^{1/2}\boldsymbol{I}\boldsymbol{X}^{1/2} = \boldsymbol{X}$ and $\boldsymbol{X}^{1/2}\boldsymbol{X}^{-1/2}\boldsymbol{Y}\boldsymbol{X}^{-1/2}\boldsymbol{X}^{1/2} = \boldsymbol{Y}$. Because $\boldsymbol{X}^{1/2}\boldsymbol{Y}\boldsymbol{X}^{1/2}$ is an isometry, the path (4.36) is a geodesic joining $\boldsymbol{X}$ and $\boldsymbol{Y}$. Thus, equality (4.32) follows. Next, in order to prove the equality (4.33) we have that,

$$\begin{aligned}
d(\boldsymbol{X}, \boldsymbol{X}\#_r \boldsymbol{Y}) &= d(\boldsymbol{X}, \boldsymbol{X}^{1/2} \left(\boldsymbol{X}^{-1/2} \boldsymbol{Y} \boldsymbol{X}^{-1/2} \right)^r \boldsymbol{X}^{1/2}), \\
&= d(\boldsymbol{X}^{-1/2} \boldsymbol{X} \boldsymbol{X}^{-1/2}, \boldsymbol{X}^{-1/2} \boldsymbol{X}^{1/2} \left(\boldsymbol{X}^{-1/2} \boldsymbol{Y} \boldsymbol{X}^{-1/2} \right)^r \boldsymbol{X}^{1/2} \boldsymbol{X}^{-1/2}), \\
&= d(\boldsymbol{I}, \left(\boldsymbol{X}^{-1/2} \boldsymbol{Y} \boldsymbol{X}^{-1/2} \right)^r), \\
&= \inf \int_0^1 \left\| \log(\boldsymbol{I}) - \log \left(\boldsymbol{X}^{-1/2} \boldsymbol{Y} \boldsymbol{X}^{-1/2} \right)^r \right\| dr, \\
&= r \inf \int_0^1 \left\| \log \left(\boldsymbol{X}^{-1/2} \boldsymbol{Y} \boldsymbol{X}^{-1/2} \right) \right\| = r\, d(\boldsymbol{X}, \boldsymbol{Y}),
\end{aligned} \tag{4.37}$$

where the last equality arises as a consequence of (4.4) since,

$$\begin{aligned}
d(\boldsymbol{X}, \boldsymbol{Y}) &= d\left(\boldsymbol{I}, \boldsymbol{X}^{-1/2} \boldsymbol{Y} \boldsymbol{X}^{-1/2} \right), \\
&= \left\| \log \boldsymbol{I} - \log \left(\boldsymbol{X}^{-1/2} \boldsymbol{Y} \boldsymbol{X}^{-1/2} \right) \right\|_{\boldsymbol{I}}, \\
&= \left\| \log \left(\boldsymbol{X}^{-1/2} \boldsymbol{Y} \boldsymbol{X}^{-1/2} \right) \right\|_{\boldsymbol{I}}.
\end{aligned} \tag{4.38}$$

Finally, from the definition of the Riemannian norm and basic linear algebra, we get that,

$$d(\boldsymbol{X}, \boldsymbol{Y}) = \left[\sum_{i=1}^{3} \left(\log \sigma_i \left(\boldsymbol{X}^{-1} \boldsymbol{Y} \right) \right)^2 \right]^{1/2}, \tag{4.39}$$

where σ_i are the eigenvalues of the matrix $\boldsymbol{X}^{-1}\boldsymbol{Y}$. □

We are now ready to introduce the scalar measures that can be used to quantify the structure tensor, $\mathcal{G}$. First, let us denote the matrix logarithm of $\mathcal{G}$ as $\mathcal{L}_{\mathcal{G}}$ (i. e., $\mathcal{G} = e^{\mathcal{L}_{\mathcal{G}}}$). This matrix logarithm exists, is unique (since $\mathcal{G}$ is positive definite) and has eigenvalues, which are the logarithm of the eigenvalues of $\mathcal{G}$.

4.5.1 *Scalar Invariant 1: Volume Ratio*

Let $\sigma_i(\mathcal{G})$ $(i = 1, 2, 3)$ be the eigenvalues of $\mathcal{G}$. Define the first scalar invariant as the volume ratio of fluctuations, δ_1, as

Definition 4.8

$$\delta_1 = \text{tr}\mathcal{L}_{\mathcal{G}} = \log\det\mathcal{G} = \log\left(\frac{\det\mathcal{C}}{\det\overline{\mathcal{C}}}\right).$$

When $\delta_1 = 0$, the mean and the instantaneous conformation tensors have the same volume and when δ_1 is negative (positive), the instantaneous conformation tensor has smaller (larger) volume than the mean volume.

4.5.2 *Scalar Invariant 2: Shortest Distance from Mean*

When $\mathcal{C} = \overline{\mathcal{C}}$, we have $\mathcal{G} = \mathbf{I}$. When $\mathcal{C} \neq \overline{\mathcal{C}}$, we wish to consider the shortest path between $\mathbf{I}$ and $\mathcal{G}$ as a measure of the magnitude of fluctuations. Using Eq. (4.39), we consider the squared geodesic distance related with this path,

Definition 4.9

$$\delta_2 = \text{tr}\mathcal{L}_{\mathcal{G}}^2 = d^2(\mathbf{I}, \mathcal{G}) = \sum_{i=1}^{3} (\log\sigma_i)^2.$$

Using Eq. (4.39), we can verify that $d^2(\mathbf{I}, \mathcal{G}) = d^2(\mathbf{I}, \mathcal{G}^{-1})$, which implies that this squared geodesic treats both expansions and compressions identically. The affine invariance property (4.4) ensures that $d(\mathbf{I}, \mathcal{G}) = d([\mathbf{I}]_{\mathbf{A}}, [\mathcal{G}]_{\mathbf{A}})$. With $\mathbf{A} = \overline{\mathbf{F}}$, we obtain,

$$d^2(\mathbf{I}, \mathcal{G}) = d^2(\overline{\mathcal{C}}, [\mathcal{G}]_{\overline{\mathcal{F}}}) = d^2(\overline{\mathcal{C}}, \mathcal{C}) = d^2(\overline{\mathcal{C}}^{-1}, \mathcal{C}^{-1}). \tag{4.40}$$

The last equality in Eq. (4.40) exhibits the fact that the metric introduced in Eq. (4.39), handles expansions and compressions on equal terms, unlike the regular Euclidean metric (or the Frobenius norm).

4.5.3 Scalar Invariant 3: Anisotropy Index

Following Hameduddin [18], we define the anisotropy index, δ_3 as the squared geodesic distance between $\mathcal{G}$ and the closest isotropic tensor,

Definition 4.10

$$\delta_3 = \inf_a d^2(a\mathbf{I}, \mathcal{G}) = \inf_a \mathrm{tr}(\mathcal{G} - (\log a)\mathbf{I})^2.$$

Through differentiation, we find that $a = \left(\prod_{i=1}^{3} \sigma_i\right)^{1/3} = (\det \mathcal{G})^{1/3}$ is the minimizing stationary point, which implies that,

$$\delta_3 = d^2((\det \mathcal{G})^{1/3}\,\mathbf{I}, \mathcal{G}) = \delta_2 - \frac{1}{3}\delta_1^2. \tag{4.41}$$

Notice that $\delta_3 = 0$ if and only if $\delta_1^2 = 3\delta_2$, in which case $\mathcal{G}$ reduces to an isotropic tensor.

Finally, we surmise that equations (4.15a), (4.15b) and (4.23) along with definitions (4.5, 4.8, 4.9 and 4.10) form the complete set of equations governing the dynamics of viscoelastic sub-diffusive flows.

4.6 Perturbative Expansion for Weakly Nonlinear Deformation

A weakly nonlinear expansion up to the Kth power of the velocity field is given by,

$$\begin{aligned} \Omega &= \overline{\Omega} + \sum_{k=1}^{K} \epsilon^k \Omega_k, \\ \psi &= \overline{\psi} + \sum_{k=1}^{K} \epsilon^k \psi_k, \end{aligned} \tag{4.42}$$

where the superscript, $\overline{(\ )}$, denote the mean values and $\Omega_k(\boldsymbol{x}, t)$, ψ_k, $(k \in [1, K])$ are the perturbed vorticities and stream-functions of the kth-order, respectively. A similar expansion for $\mathcal{G}$ is inappropriate because it is positive definite, and there is no *a priori* guarantee on this property with regular arithmetic expansion. Instead, we adopt the geometric expansion by multiplicatively decomposing the fluctuating deformation gradient tensor into K separate components,

$$\mathcal{L}_{\text{Pert}} = \mathcal{L}_1\mathcal{L}_2 \ldots \mathcal{L}_k, \tag{4.43}$$

Using (4.5) as well as the matrix logarithm of $\mathcal{G}$ ($=\mathcal{L}_{\mathcal{G}}$), we can express the perturbed Structure Tensor at kth-stage of decomposition as,

$$\mathcal{G}_k^{\epsilon^k} = \mathcal{L}_k \mathcal{L}_k^T = e^{\epsilon^k \mathcal{L}_{\mathcal{G}_k}}, \tag{4.44}$$

where $\mathcal{L}_{\mathcal{G}_0} = \mathbf{0}$. From Eq. (4.43), we can associate a perturbed tensor,

$$\mathcal{L}_k = e^{\epsilon^k \mathcal{L}_{\mathcal{G}_k}/2}. \tag{4.45}$$

At this stage, we remark that although we assume that each $\mathcal{L}_k$ is positive definite, the product of positive definite tensors, $\mathcal{L}_{\text{Pert}}$ (equation (4.43)) is not necessarily positive definite. However, since $\mathcal{L}_{\text{Pert}}\mathcal{L}_{\text{Pert}}^T = \mathcal{L}\mathcal{L}^T = \mathcal{G}$, we can show via a polar decomposition that $\mathcal{L}_{\text{Pert}} = \mathcal{L}\boldsymbol{R}$ for some rotation tensor, $\boldsymbol{R}$. Substituting equation (4.45) and (4.43) in equation (4.44), we arrive at the necessary expansion,

$$\begin{aligned}
\mathcal{G} &= e^{\epsilon \mathcal{L}_{\mathcal{G}_1}/2} \dots e^{\epsilon^{K-1} \mathcal{L}_{\mathcal{G}_{K-1}}/2} e^{\epsilon^K \mathcal{L}_{\mathcal{G}_K}} e^{\epsilon^{K-1} \mathcal{L}_{\mathcal{G}_{K-1}}/2} \dots e^{\epsilon \mathcal{L}_{\mathcal{G}_1}/2}, \\
&= \boldsymbol{I} + \epsilon \mathcal{L}_{\mathcal{G}_1} \\
&+ \epsilon^2 \left(\frac{\mathcal{L}_{\mathcal{G}_1}^2}{2} + \mathcal{L}_{\mathcal{G}_2} \right) + \epsilon^3 \left(\frac{\mathcal{L}_{\mathcal{G}_1}^3}{6} + \frac{1}{2}(\mathcal{L}_{\mathcal{G}_1}\mathcal{L}_{\mathcal{G}_2} + (\mathcal{L}_{\mathcal{G}_1}\mathcal{L}_{\mathcal{G}_2})^T) + \mathcal{L}_{\mathcal{G}_3} \right) + \dots,
\end{aligned} \tag{4.46}$$

where the second equality in equation (4.46) makes use of the matrix exponential, $e^{\epsilon^k \mathcal{L}_{\mathcal{G}_k}} = \sum_{q=0}^{\infty} \epsilon^{kq} \mathcal{L}_{\mathcal{G}_k}^q / q!$. Substituting expansions (4.42) and (4.46) in equations (4.15a), (4.15b) and (4.23), we arrive at the $\mathcal{O}(\epsilon)$ model equations,

$$Re\left[\frac{\partial^\alpha \Omega_1}{\partial t^\alpha} + \overline{\mathbf{v}} \cdot \nabla \Omega_1 + \mathbf{v}_1 \cdot \nabla \overline{\Omega}\right] = \nu \nabla^2 \Omega_1 + \frac{1-\nu}{W_e} \nabla \times \nabla \cdot \left(\overline{\boldsymbol{F}} \mathcal{L}_{\mathcal{G}_1} \overline{\boldsymbol{F}}^T\right), \tag{4.47a}$$

$$\nabla^2 \psi_1 = -\Omega_1, \tag{4.47b}$$

$$\frac{\partial^\alpha}{\partial t^\alpha} \mathcal{L}_{\mathcal{G}_1} + \overline{\mathbf{v}} \cdot \nabla \mathcal{L}_{\mathcal{G}_1} = 2\text{sym}\left(\mathfrak{F}(\mathbf{v}_1) + \mathcal{L}_{\mathcal{G}_1} \mathfrak{F}(\overline{\mathbf{v}})\right) - \frac{1}{W_e} \mathcal{L}_{\mathcal{G}_1}, \tag{4.47c}$$

as well as the $\mathcal{O}(\epsilon^2)$ model equations,

$$Re\left[\frac{\partial^{\alpha}\Omega_2}{\partial t^{\alpha}}+\overline{\mathbf{v}}\cdot\nabla\Omega_2+\mathbf{v}_2\cdot\nabla\overline{\Omega}+\mathbf{v}_1\cdot\nabla\Omega_1\right]=\nu\nabla^2\Omega_2+\frac{1-\nu}{W_e}\nabla\times\nabla\cdot\left(\overline{\boldsymbol{F}}\left(\frac{\mathcal{L}_{\mathcal{G}_1}^2}{2}+\mathcal{L}_{\mathcal{G}_2}\right)\overline{\boldsymbol{F}}^T\right), \tag{4.48a}$$

$$\nabla^2\psi_2=-\Omega_2, \tag{4.48b}$$

$$\begin{aligned}\frac{\partial^{\alpha}\mathcal{L}_{\mathcal{G}_2}}{\partial t^{\alpha}}+\overline{\mathbf{v}}\cdot\nabla\mathcal{L}_{\mathcal{G}_2}+\mathbf{v}_1\cdot\nabla\mathcal{L}_{\mathcal{G}_1}&=2\,\mathrm{sym}\left(\mathfrak{F}(\mathbf{v}_2)+\mathcal{L}_{\mathcal{G}_2}\mathfrak{F}(\overline{\mathbf{v}})\right)+\frac{\mathcal{L}_{\mathcal{G}_1}^2}{2W_e}-\frac{\mathcal{L}_{\mathcal{G}_2}}{W_e}\\&-\mathcal{L}_{\mathcal{G}_1}\mathrm{sym}\left(\mathfrak{F}(\overline{\mathbf{v}})\right)\mathcal{L}_{\mathcal{G}_1}+\mathcal{L}_{\mathcal{G}_1}\mathrm{asym}\left(\mathfrak{F}(\mathbf{v}_1)\right)\\&-\mathrm{asym}\left(\mathfrak{F}(\mathbf{v}_1)\right)\mathcal{L}_{\mathcal{G}_1},\end{aligned} \tag{4.48c}$$

where $\mathrm{sym}(\boldsymbol{A})=(\boldsymbol{A}+\boldsymbol{A}^T)/2$, $\mathrm{asym}(\boldsymbol{A})=(\boldsymbol{A}-\boldsymbol{A}^T)/2$ and $\mathbf{v}_i=(u_i,v_i)=(\frac{\partial\psi_i}{\partial y},-\frac{\partial\psi_i}{\partial x})$.

Linear Perturbations

As an illustration, we highlight the case of linear perturbative solutions for the 2D viscoelastic channel flow for polymer melts. A rectilinear coordinate system is used with x, y denoting the channel flow direction and the transverse direction, respectively. The origin of this coordinate system is chosen at the left end of the lower wall of the channel. The size of the domain is chosen to be $(x, y)\in\Gamma=[0,\ 5]\times[0,\ 1]$. The mean flow is assumed to be a plane Poiseuille flow with its variation entirely in the transverse direction, namely,

$$\mathbf{U}_0=(y-y^2)\mathbf{e_x}, \tag{4.49}$$

where $\mathbf{e_x}$ is the unit vector along x-direction. The mean flow, $\mathbf{U}_0$, defines the mean vorticity, $\overline{\Omega}=2y-1$, and the mean stream-function, $\overline{\psi}=\frac{y^2}{2}-\frac{y^3}{3}$. In this case, the initial conditions for the perturbed solution can be constructed via the superposition of the mean flow and the instability mode, as follows,

$$\begin{aligned}\Omega|_{t=0}&\approx\overline{\Omega}+\epsilon\Omega_1=\overline{\Omega}+\epsilon\{\widetilde{\Omega}(y)|_{t=0}e^{ikx}\}_r,\\ \psi|_{t=0}&\approx\overline{\psi}+\epsilon\psi_1=\overline{\psi}+\epsilon\{\widetilde{\psi}(y)|_{t=0}e^{ikx}\}_r,\\ \mathcal{G}|_{t=0}&\approx\boldsymbol{I}+\epsilon\mathcal{L}_{\mathcal{G}_1}=\boldsymbol{I}+\epsilon\{\widetilde{\mathcal{L}_{\mathcal{G}}}(y)|_{t=0}e^{ikx}\}_r,\end{aligned} \tag{4.50}$$

where $(\Omega_1,\psi_1,\mathcal{L}_{\mathcal{G}_1})$ are the perturbations that are Fourier transformed in the x-direction. $\mathcal{R}\{\}$ denotes the real part of the complex valued function. The equations governing the initial conditions (4.50) are listed below. Assuming a normal mode expansion for the perturbed field, $\phi_1=\widetilde{\phi(y)}e^{ikx}$ (where $\boldsymbol{\phi}_1=(\Omega_1,\psi_1,\mathcal{L}_{\mathcal{G}_1})$), Eq. (4.47) reduces to,

$$
\begin{aligned}
&Re\left((y-y^2)\widetilde{\Omega}(ik)-2\widetilde{\psi}(ik)\right)=\nu\left(\widetilde{\Omega}(ik)^2+\widetilde{\Omega}''\right)+\frac{1-\nu}{We}\Big(-k^2F_1F_2\widetilde{\mathcal{L}\mathcal{G}_{11}}-k^2F_2^2\\
&\widetilde{\mathcal{L}\mathcal{G}_{12}}-k^2F_1F_4\widetilde{\mathcal{L}\mathcal{G}_{12}}-k^2F_2F_4\widetilde{\mathcal{L}\mathcal{G}_{22}}-F_1''F_2\widetilde{\mathcal{L}\mathcal{G}_{11}}-F_1F_2''\widetilde{\mathcal{L}\mathcal{G}_{11}}-F_1F_2\widetilde{\mathcal{L}\mathcal{G}_{11}}''-F_2^{2''}\\
&\widetilde{\mathcal{L}\mathcal{G}_{12}}-F_2^2\widetilde{\mathcal{L}\mathcal{G}_{12}}''-F_1''F_4\widetilde{\mathcal{L}\mathcal{G}_{12}}-F_1F_4''\widetilde{\mathcal{L}\mathcal{G}_{12}}-F_1F_4\widetilde{\mathcal{L}\mathcal{G}_{12}}''-F_2''F_4\widetilde{\mathcal{L}\mathcal{G}_{22}}-F_2F_4''\widetilde{\mathcal{L}\mathcal{G}_{22}}\\
&-F_2F_4\widetilde{\mathcal{L}\mathcal{G}_{22}}''-2F_1'F_2'\widetilde{\mathcal{L}\mathcal{G}_{11}}-2F_1'F_2\widetilde{\mathcal{L}\mathcal{G}_{11}}'-2F_1'F_4'\widetilde{\mathcal{L}\mathcal{G}_{12}}-2F_1'F_4\widetilde{\mathcal{L}\mathcal{G}_{12}}'-2F_1F_2'\\
&\widetilde{\mathcal{L}\mathcal{G}_{11}}'-2F_1F_4'\widetilde{\mathcal{L}\mathcal{G}_{12}}'-2F_2'F_4'\widetilde{\mathcal{L}\mathcal{G}_{22}}-2F_2'F_4\widetilde{\mathcal{L}\mathcal{G}_{22}}'-4F_2F_2'\widetilde{\mathcal{L}\mathcal{G}_{12}}'-2F_2F_4'\widetilde{\mathcal{L}\mathcal{G}_{22}}'+\\
&2ikF_2F_2'\widetilde{\mathcal{L}\mathcal{G}_{11}}+ikF_2^2\widetilde{\mathcal{L}\mathcal{G}_{11}}'+2ikF_2'F_4\widetilde{\mathcal{L}\mathcal{G}_{12}}+2ikF_2F_4'\widetilde{\mathcal{L}\mathcal{G}_{12}}+2ikF_2F_4\widetilde{\mathcal{L}\mathcal{G}_{12}}'+2ik\\
&F_4F_4'\widetilde{\mathcal{L}\mathcal{G}_{22}}+ikF_4^2\widetilde{\mathcal{L}\mathcal{G}_{22}}'-2ikF_1F_1'\widetilde{\mathcal{L}\mathcal{G}_{11}}-ikF_1^2\widetilde{\mathcal{L}\mathcal{G}_{11}}'-2ikF_1'F_2\widetilde{\mathcal{L}\mathcal{G}_{12}}-2ikF_1F_2'\\
&\widetilde{\mathcal{L}\mathcal{G}_{12}}-2ikF_1F_2\widetilde{\mathcal{L}\mathcal{G}_{12}}-2ikF_2F_2'\widetilde{\mathcal{L}\mathcal{G}_{22}}-ikF_2^2\widetilde{\mathcal{L}\mathcal{G}_{22}}'\Big),
\end{aligned}
\tag{4.51a}
$$

$$
\widetilde{\psi}(ik)^2+\widetilde{\psi}''=-\widetilde{\Omega},
\tag{4.51b}
$$

$$
\begin{aligned}
&(y-y^2)(ik)\widetilde{\mathcal{L}\mathcal{G}_{11}}=\frac{2}{\sqrt{d}}\Big(F_1F_4\widetilde{\psi}'(ik)-F_1F_2\widetilde{\psi}''-F_2F_4\widetilde{\psi}(ik)^2+F_2^2\widetilde{\psi}'(ik)-\\
&\widetilde{\mathcal{L}\mathcal{G}_{11}}F_1F_2(1-2y)-F_2^2\widetilde{\mathcal{L}\mathcal{G}_{12}}(1-2y)-\widetilde{\psi}(ik)\left(-F_4F_1'+F_2F_2'\right)\Big)-\frac{\widetilde{\mathcal{L}\mathcal{G}_{11}}}{We},
\end{aligned}
\tag{4.51c}
$$

$$
\begin{aligned}
&(y-y^2)(ik)\widetilde{\mathcal{L}\mathcal{G}_{12}}=\frac{1}{\sqrt{d}}\Big(-2F_1F_2\widetilde{\psi}'(ik)+F_1^2\widetilde{\psi}''+F_2^2\widetilde{\psi}(ik)^2+\widetilde{\mathcal{L}\mathcal{G}_{11}}F_1^2(1-2y)+\\
&2F_2F_4\widetilde{\psi}'(ik)-F_2^2\widetilde{\psi}''-F_4^2\widetilde{\psi}(ik)^2-\widetilde{\mathcal{L}\mathcal{G}_{22}}F_2^2(1-2y)-\widetilde{\psi}(ik)\left(F_2F_1'-F_1F_2'-F_4\right.\\
&\left.F_2'+F_2F_4'\right)\Big)-\frac{\widetilde{\mathcal{L}\mathcal{G}_{12}}}{We},
\end{aligned}
\tag{4.51d}
$$

$$
\begin{aligned}
&(y-y^2)(ik)\widetilde{\mathcal{L}\mathcal{G}_{22}}=\frac{2}{\sqrt{d}}\Big(\widetilde{\mathcal{L}\mathcal{G}_{12}}F_1^2(1-2y)-F_2^2\widetilde{\psi}'(ik)+F_1F_2\widetilde{\psi}''+F_2F_4\widetilde{\psi}(ik)^2-\\
&F_1F_4\widetilde{\psi}'(ik)+\widetilde{\mathcal{L}\mathcal{G}_{22}}F_1F_2(1-2y)-\widetilde{\psi}(ik)\left(F_2F_2'-F_1F_4'\right)\Big)-\frac{\widetilde{\mathcal{L}\mathcal{G}_{22}}}{We},
\end{aligned}
\tag{4.51e}
$$

where $\dfrac{d()}{dy}=(\ )'$ and,

$$
\overline{\boldsymbol{F}}=\begin{bmatrix}F_1 & F_2\\ F_2 & F_4\end{bmatrix}=\begin{bmatrix}\dfrac{1+\sqrt{d}}{\sqrt{2d+2\sqrt{d}}} & \dfrac{We(1-2y)}{\sqrt{2d+2\sqrt{d}}}\\[2ex] \dfrac{We(1-2y)}{\sqrt{2d+2\sqrt{d}}} & \dfrac{2d+\sqrt{d}-1}{\sqrt{2d+2\sqrt{d}}}\end{bmatrix},
$$

where

$$
d=1+We^2(1-2y)^2.
$$

The solution to the boundary value problem (4.51) is found subject to the boundary conditions, $(\widetilde{\psi}(y), \widetilde{\psi}'(y)) = (0, 0)$ at the rigid walls $y = 0, 1$, which is unique up to an integration constant for the stream-function perturbation (refer Fig. 4.1b for the solution).

An initial condition comprising a small amplitude unstable mode will initially grow exponentially, as predicted by the linear theory. However, nonlinear effects eventually become significant since the conformation tensor loses positive definiteness otherwise. Hence, our interest in studying linear perturbations is to find an estimate of the maximum time during which the perturbed solution can be well approximated by the linear theory, i.e., along the Euclidean manifold.

Consider an initial condition which is a perturbed base flow, $\mathcal{G}|_{t=0} = \boldsymbol{I} + \epsilon \mathcal{L}_{\mathcal{G}}$. If we assume that the perturbed mode grows according to linear theory for some time and $\mathcal{G}$ evolves along Euclidean lines with growth rate $\omega > 0$, then $\mathcal{G}(t) = \boldsymbol{I} + \epsilon \mathcal{L}_{\mathcal{G}} e^{\omega t}$. Suppose $\mathcal{L}_{\mathcal{G}}$ is not zero and is harmonic in the spatial direction, then $\mathcal{L}_{\mathcal{G}}$ has a strictly negative eigenvalue somewhere in the domain. This is because a harmonic perturbation leads to regions where the polymers are much more compressed than the maximum expansion, in a volumetric sense, since positive and negative additive perturbations to the mean conformation tensor, $\overline{\mathcal{C}}$, with equal magnitude are not of equal magnitude with respect to the natural norm on $\boldsymbol{PS}_3$. For positive-definiteness of $\mathcal{G}$, we require the eigenvalues, $1 + \epsilon \sigma_i(\mathcal{L}_{\mathcal{G}}) e^{\omega t} > 0, \ (i = 1, 2, 3)$. Whenever $\sigma_i(\mathcal{L}_{\mathcal{G}}) < 0$, the dynamics induces a curvature on the evolution of the perturbed mode along $\boldsymbol{PS}_3$ before the time, t_m, when the eigenvalue of $\mathcal{G}$ crosses zero. Hence, at the point of crossing zero, this time is given by,

$$\omega t_m = -\left(\log \epsilon + \log \max |\sigma_i(\mathcal{L}_{\mathcal{G}})|\right), \tag{4.52}$$

where $\max |\sigma_i(\mathcal{L}_{\mathcal{G}})|$ is the magnitude of the largest negative eigenvalue in the domain. Refer to the previous chapter on the instability of viscoelastic sub-diffusive channel flows, by the authors [19] to gain an insight on the procedure for finding, $\sigma_i(\mathcal{L}_{\mathcal{G}})$. Equation (4.52) serves as a guide for selecting the initial perturbation amplitude, ϵ, based on the time, t_m, i e., by reducing ϵ we can arbitrarily increase t_m to a desired value.

In order to compare the linear evolution of the unstable modes (Eq. (4.47) with initial conditions (4.50)) with the nonlinear modes (Eqs. (4.15a), (4.15b) and (4.23)), the following quantity is utilized,

$$A(t) = \frac{1}{\ell_x \ell_y} \int_{\Gamma} d^2(\boldsymbol{I}, \mathcal{G}) d\Gamma, \tag{4.53}$$

which measures the perturbations away from the isotropic tensor, $\boldsymbol{I}$, in the volume-averaged sense [18]. (ℓ_x, ℓ_y) are the lengths of the domain in the flow direction and the transverse direction, respectively. The time evolution of the normalized function, $A(t)/A(0)$, for parameter values, $We = 10.0$, $Re = 70.0$, $\nu = 0.3$, $\alpha = 0.5$, $\epsilon = 0.1$, $k = 0.01$ (refer Eqs. (4.15a), (4.15b) and (4.23) as well as the initial conditions (4.50)) is shown in Fig. 4.1b. We note that the evolution of $A(t)/A(0)$

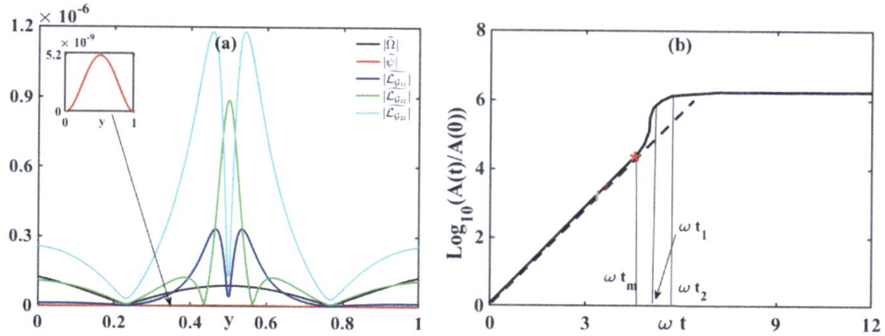

Fig. 4.1 **a** Solution to the linearized system of equations (4.51) subject to the boundary values, $(\widetilde{\psi}(y), \widetilde{\psi}'(y)) = (0, 0)$ at the rigid channel walls, $y = 0$ and $y = 1$, and **b** time evolution of A, as defined in equation (4.53) for parameter values, $We = 10.0$, $Re = 70.0$, $\nu = 0.3$ and $\alpha = 0.5$. The solid line represents the nonlinear evolution, $A(t)/A(0)$ (Eqs. ((4.15a), (4.15b) and (4.23)) with initial conditions (4.50)) while the dashed line represents the growth of the solution predicted by the linear theory (4.47). The asterisk (∗) indicates the maximum time, t_m as defined in equation (4.52)

matches with the one predicted by the linear theory (4.47), up to the maximum time, $\omega t_m \approx 4.6$ and then shows deviation from the linear growth in the form of an exponential growth upto time, $\omega t_1 \approx 5.2$, followed by an eventual saturation at $\omega t_2 \approx 6.0$. While the initial exponential deviation can be explained due to the exponential form (refer equation (4.46)) of the Structure Tensor, the mode saturation is the manifestation of nonlinear effects, which is absent in the linear theory.

4.7 Conclusion

In this chapter, we have developed a mathematically consistent decomposition of the conformation tensor, $\mathcal{C}$, into the Structure Tensor, $\mathcal{G}$ (4.5), for viscoelastic subdiffusive flows that resolve the difficulties associated with the traditional arithmetic decomposition. We characterized the fluctuations in $\mathcal{G}$ by using a geometry specifically constructed for $\boldsymbol{PS}_3$ and obtained three scalar measures: the volume ratio, δ_1 (4.8), the shortest distance from the mean conformation, δ_2 (4.9), and the anisotropy index, δ_3 (4.10). The linear perturbation studies and the fully nonlinear simulations provided interesting insights about the instantaneous polymer conformation tensor that is not readily available from an arithmetic decomposition of $\mathcal{C}$, including (i) evaluation of a (perturbation amplitude-dependent) maximum time during which the linear perturbative solution can be well approximated by the weakly nonlinear solution, along the Euclidean manifold, (ii) a better measure to detect neighbourhoods where the mean conformation tensor tends to be significantly different in comparison to the instantaneous conformation tensor.

While the analysis presented here has delivered a general framework to provide a quantitative explanation of the previously published experimental findings [1–4], the detailed physics of the flow-induced structure formation in viscoelastic sub-diffusive flows is given in the next chapter.

4.8 Problems

1. Let $M = \mathbb{R}^d$ with its standard C^∞ manifold structure. Give an example of a complete Riemannian metric on M, and also an example of one non-complete Riemannian metric on M.
2. Let N be a complete Riemannian manifold and let M be an embedded submanifold of N which is closed. Prove that M is complete.
3. [Properties useful in the proof of Hopf-Rinow Theorem] Let (X, d) be an arbitrary metric space. Recall that for $p \in X$ and $r > 0$ we write $B_r(p)$ for the open ball, $B_r(p) = \{q \in X | d(p, q) < r\}$.

   a. Prove that d is a continuous function $(X \times X \to \mathbb{R}_{\geq 0})$.
   b. Prove that for any $p \in X, r > 0$,

   $$\partial B_r(p) \subset \{q \in X | d(p, q) = r\},$$

   and both these sets are closed. Furthermore if (X, d) is a Riemannian manifold then equality holds: $\partial B_r(p) = \{q \in X | d(p, q) = r\}$.
   c. Continue to assume that (X, d) is a Riemannian manifold. Let $p, q \in X, r > 0$, and assume $d(p, q) > r$. Assume that p_0 is a point on $\partial B_r(p)$ where $d(., q)_{\partial B_r(p)}$ is minimal. Prove that $d(p, q) = d(p, p_0) + d(p_0, q)$.

References

1. Riley JJ, Hak MG, Metcalfe RW (1988) Complaint coatings. Ann Rev Fluid Mech 20:393–403
2. Nandagopalan P, John J, Baek SW, Miglani A, Ardhianto K (2018) Shear-flow rheology and viscoelastic instabilities of ethanol gel fuels. Exp Thermal Fluid Sci 99:181–189
3. Zarabadi M (2019) Development of a Robust microfluidic electrochemical cell for biofilm study in controlled hydrodynamic conditions. University Laval
4. Zarabadi MP, Charette SJ, Greener J (2018) Flow-based deacidification of geobacter sulfurreducens biofilms depends on nutrient conditions: a microfluidic bioelectrochemical study. Chem Electrochem 5(23):3645–3653
5. Khalid M, Chaudhary I, Garg P, Shankar V, Subramanian G (2021) The centre-mode instability of viscoelastic plane Poiseuille flow. J Fluid Mech 915:A43
6. Dubief Y, Terrapon VE, White CM, Shaqfeh ESG, Moin P, Lele SK (2005) New answers on the interaction between polymers and vortices in turbulent flows. Turbul Combust Former Appl Sci Res 74(4):311–329

7. Kim K, Sureshkumar R (2013) Spatiotemporal evolution of hairpin eddies. Reynolds stress and polymer torque in drag-reduced turbulent channel flows. Phys Rev E 87(6):063002
8. Bird R, Armstrong R, Hassager O (1987) Dynamics of polymeric liquids. Wiley
9. Sureshkumar R, Beris AN, Handler R (1997) Direct numerical simulation of the turbulent channel flow of a polymer solution. Phys Fluids 9(3):743–755
10. Beris AN, Edwards BJ (1994) Thermodynamics of flowing systems: with internal microstructure. Oxford University Press
11. Bhatia R (2015) Positive definite matrices. Princeton University Press
12. Fattal R, Kupferman R (2004) Constitutive laws for the matrix-logarithm of the conformation tensor. J non-Newt Fluid Mech 123(2):281–285
13. Lang S (2001) Fundamentals of differential geometry, graduate texts in mathematics. Springer, p 191
14. Podlubny I (1999) Fractional differential equations. Academic Press, p 198
15. Lovett S (2019) Differential geometry of manifolds. CRC Press
16. Lee J (2019) Introduction to Riemannian manifold, graduate texts in mathematics. Springer, vol 176
17. Sircar S, Bansal D (2019) Spatiotemporal linear stability of viscoelastic free shear flows: dilute regime. Phys Fluids 31(8):084104
18. Hameduddin I (2018) Tackling viscoelastic turbulence. John Hopkins University
19. Chauhan T, Bansal D, Sircar S (2023) Spatiotemporal linear stability of viscoelastic subdiffusive channel flows: a fractional calculus framework. J Eng Math 141(1):8

Chapter 5
Rheodynamics of Sub-diffusive Channel Flows

Abstract Direct numerical simulations of viscoelastic, sub-diffusive, plane Poiseuille flow, representing thick polymer solutions including polymer melts, flows of liquid crystals, as well as biological flow such as mucus, are described. The computations are carried out using a recently developed family of time-asymptotically stable, implicit-explicit, adaptive, time integration methods (denoted with the θ-method) applicable for the general solution of the fractional advection-diffusion-reaction (FADR) equations. The spectral analysis of the method (involving the group velocity and the phase speed) indicates a region of favourable dispersion for a limited range of Peclet number. The accuracy and the efficacy of the method are benchmarked using the two-dimensional fractional diffusion equation, originally proposed by Brunner (J. Comput. Phys. 229 6613-6622 (2010)). Numerical simulations of the viscoelastic channel flow effectively capture the non-homogeneous regions of high viscosity at moderate to high fluid inertia (or the so-called 'spatiotemporal macrostructures'), experimentally observed in the flow-instability transition of sub-diffusive flows.

5.1 Introduction

The subject of anomalous diffusion has received tremendous attention over the last half-century, ranging from physics [1], biology [2] to quantitative finance [3]. Some of the most enigmatic and profoundly significant experimental results are better rationalized within the viscoelastic sub-diffusive approach in random environments such as the cytosol and the plasma membrane of biological cells [4], crowded complex fluids and polymer solutions [5], dense colloidal suspensions [6], single-file diffusion in colloidal systems [7] as well as in atherosclerotic blood vessels [8]. The observed (anomalous) sub-diffusion often combines features of ergodic fractional Brownian motion (reflecting viscoelasticity) and the nonergodic jumplike non-Markovian diffusional processes (reflecting disorder) [9, 10]. The sub-diffusive object is considered primarily as being elastic and structurally robust, although it requires 'fluidity' and flexibility besides its elasticity for proper functioning, e.g., consider a viscoelastic nanoscaled polymer drop armed with a rigid backbone that can take on different macroscopic conformations [11].

H. Wilson et al., *Viscoelastic Subdiffusive Flows*, Fluid Mechanics and Its Applications 138, https://doi.org/10.1007/978-981-96-0135-6_5

One objective of this chapter is to highlight the potential of fractional calculus to effectively highlight the dynamics of viscoelastic sub-diffusive flows. Fractional calculus serves as a powerful tool for modelling the constitutive relations in the linear [12] as well as nonlinear viscoelasticity theory [6] and to explain certain paradoxical experimental findings [1] (one such experimentally abnormal feature captured by our model, namely the occurrence of temporal stability at high fluid inertia, was identified in Chap. 2 and details are further presented in Sect. 5.5.4). Gemant highlighted the relaxation curves for some viscoelastic fluids by employing a fractional viscoelastic model for the first time [12]. Scott-Blair developed a new constitutive law (known as the 'fractional Newton model') to describe the experimental outcome of Gemant on stress relaxation [13]. Caputo introduced the fractional Voigt model to simulate the dissipation in seismology [14]. Bagley and Torvik [15] showed that there exists a quantitative connection between the fractional viscoelastic model (at the macroscopic level) and the molecular theory of Rouse's polymer chain melts [16]. With the development of the fractional viscoelastic model, the flow of the fractional viscoelastic fluid has been extensively investigated [17–21]. Tan and Xu [17] used Laplace transforms to obtain the analytical solution for velocity and stress of the plane surface flow of a fractional Maxwell fluid. Qi and Xu [18] studied the plane Poiseuille flow and plane Couette flow of a generalized Oldroyd-B fluid with fractional derivative. Zheng et al. [20] found the analytical solution for the velocity and stress of magnetohydrodynamic flow of a generalized Oldroyd-B fluid generated with an accelerating plate with fractional derivatives. Recently, Zhao et al. [21] considered the natural convection heat transfer of viscoelastic fluid with fractional Maxwell model over a vertical plate. More recent applications of the fractional viscoelastic flows include the study of the stability of coastal morphodynamics and seafloor topology [22], regulation of tissue morphodynamics [23], and capturing spatiotemporal disorder in anomalous transport of viscous flows [24].

The detailed exploration of the existing literature serves as a clear motivation for the work reported here, which is to provide a comprehensive picture of the nonlinear stability of the two-dimensional (2D), viscoelastic, sub-diffusive, fully developed Poiseuille flow through numerical simulations. The present work significantly differs from the existing in silico studies of the classical (or integer order) viscoelastic channel flows [25]), in the sense that we analyse the role of the fractional order of the time derivative, α, in the destabilization of the flow. This chapter is organized as follows. Section 5.2 describes a recently developed numerical method applicable to the general class of 2D fractional advection-reaction-diffusion (FADR) equations. Using the 1D linear FADR equation, the time-asymptotic stability analysis and the spectral analysis for the method are outlined in Sect. 5.3. The numerical method is benchmarked using the test cases for the 2D fractional diffusion equation proposed by Brunner [26], in Sect. 5.4. The numerical results of the sub-diffusive dynamics of the viscoelastic channel flow are outlined in Sect. 5.5. Section 5.6 concludes with a brief discussion of the implication of these results in future studies.

5.2 Adaptive Implicit-Explicit Time Integration Method

A growing number of applications of fractional derivatives in various fields of science and engineering indicates that there is a significant demand for better numerical algorithms. However, such algorithms are predominantly designed for one-dimensional problems [27] due to the severe memory restrictions imposed by these derivatives, a challenge which we alleviate through time adaptivity, as outlined in Sect. 5.2.2. The individual physics or scale components in fractional partial differential equations have very different properties that are reflected in their discretization; for example, in Fractional Advection Diffusion Reaction (FADR) systems [28], the discrete advection has a relatively slow (or 'nonstiff') dynamics while the diffusion is a fast (or 'stiff') evolving process. Implicit-Explicit integrators have been proposed as an attractive alternative (compared with the fully explicit or fully implicit time integration methods) where one combines the explicit (implicit) integration for the slow (fast) scale [29]. In an Implicit-Explicit scheme, the system of equations assumes the form [30],

$$\frac{\partial^{\alpha}\Phi}{\partial t^{\alpha}} = \mathbf{f}(\Phi) + \eta\mathbf{g}(\Phi), \tag{5.1}$$

where the superscript, α, denotes order of the fractional derivative and η is a non-negative parameter. In Eq. (5.1), the terms collected in $\mathbf{f}(\Phi)$ are on a slow time scale. Because they are (possibly) nonlinear, the implementation of a fully implicit integration scheme faces performance challenges, either from poor performance of iterative solvers or from a complex Jacobian matrix associated with the problem. Therefore, it makes sense to solve this term explicitly. The terms in $\mathbf{g}(\Phi)$, however, are on a fast time scale, and their explicit solution may require excessively small time steps in order to maintain numerical stability. Being usually linear in nature, they can be solved implicitly without further complications. The stability of such methods is still bounded by the Courant-Friedrichs-Lewy (CFL) condition, but because the fast terms are treated implicitly, these conditions are less strict when compared to a fully explicit scheme of similar formal order of accuracy.

We have limited our focus on the development, analysis and applicability of a novel family of spatiotemporal discretization methods for the numerical solution of the 1D and 2D FADR systems, since the model for sub-diffusive viscoelastic channel flows is a subset of the FADR system. In our present knowledge, a detailed analysis of the numerical methods for the FADR equation in the finite difference framework is absent. Such analysis is available for the fraction diffusion equation [26] and the advection-diffusion equation [31]. We describe this numerical method, next.

Let Γ be a bounded domain in $\mathbb{R}^2$ with sufficiently smooth boundary $\partial\Gamma$. The model equations for an initial-boundary value problem for the time-dependent FADR equation with fractional time-derivative of order $\alpha \in (0, 1)$ are as follows,

$$\begin{aligned}
&\frac{\partial^{\alpha} u(\mathbf{x}, t)}{\partial t^{\alpha}}+K_1(\mathbf{x}, t)\nabla \cdot u(\mathbf{x}, t)=K_2(\mathbf{x}, t)\nabla^2 u(\mathbf{x}, t)+f(\mathbf{x}, t), \ \ \mathbf{x} \in \Gamma, \ t \in (0, T),\\
&u(\mathbf{x}, 0) = u_0(\mathbf{x}), \quad \mathbf{x} \in \Gamma,\\
&\alpha_1 u(\mathbf{x}, t) + \alpha_2 \frac{\partial u(\mathbf{x}, t)}{\partial n} = g_0(\mathbf{x}, t), \quad \mathbf{x} \in \partial\Gamma, \quad t \in (0, T),
\end{aligned} \tag{5.2}$$

where $\frac{\partial^{\alpha} u(\mathbf{x},t)}{\partial t^{\alpha}}$ denotes the Caputo fractional derivative of order α [32] with respect to t defined by

$$\frac{\partial^{\alpha} u(\mathbf{x}, t)}{\partial t^{\alpha}} = \frac{1}{\Gamma(1-\alpha)} \int_0^t \frac{dt'}{(t-t')^{\alpha}} \frac{\partial u(\mathbf{x}, t')}{\partial t'}, \quad 0 < \alpha < 1, \tag{5.3}$$

and the operators $\nabla(\cdot)$ and $\nabla^2(\cdot)$ in equation (5.2), are the (integer order) gradient and the Laplacian operator in $\mathbb{R}^2$. The FADR equation is related to the non-Markovian continuous-time random walk and is a model for anomalous diffusional flow fields such as polymer melts [33], flows of liquid crystals [34] as well as biological flows including mucus [35] and cartilage [36].

5.2.1 *Time Integration*

The time integration is carried out using the standard L1 method for Caputo derivatives (refer Eq. (2.31)). Following an earlier work which utilized a standard integer-order advection-reaction-diffusion (ADR) equations [37], we retain the 'Implicit-Explicit method philosophy' by explicitly discretizing the advection and the reaction term (i. e., '$K_1 \nabla \cdot u(\mathbf{x}, t)$' term and '$f(\mathbf{x}, t)$' term, respectively in equation (5.2)), while the diffusive term (i. e., '$K_2(\mathbf{x}, t)\nabla^2 u(\mathbf{x}, t)$' term in Eq. (5.2)) is treated semi-implicitly, as follows,

$$K_2(\mathbf{x}, t)\nabla^2 u(\mathbf{x}, t) \approx \theta K_2(\mathbf{x}, t)\nabla^2 u(\mathbf{x}, t)|_{t_k} + (1-\theta) K_2(\mathbf{x}, t)\nabla^2 u(\mathbf{x}, t)|_{t_{k-1}}. \tag{5.4}$$

This family of time integration method (referred to as the θ-method in subsequent discussion) generalizes the computationally explicit L1-method by Brunner [26] as well as the fully implicit method recently proposed by Jannelli [28]. The next two sections delve into the issues of time adaptivity and spatial discretization.

5.2.2 *Adaptive Time Stepping*

When the simulation time is long, the size of memory in the fractional derivative approximation becomes enormously large. However, according to the fading memory property [38] for long times, the solution of the FADR systems changes more slowly than the standard integer-order ADR processes. Hence, it makes sense to employ a large time step at longer times. Let $\widetilde{u(\cdot, t_k)}$ be the numerical approximation for $u(\cdot, t_k)$. To detect this change, we define a measure between the numerical solutions of two consecutive time steps by,

$$\Delta u_{t_k} = \frac{\|\widetilde{u(\cdot, t_k)} - \widetilde{u(\cdot, t_{k-1})}\|_{l^2}}{\|\widetilde{u(\cdot, t_{k-1})}\|_{l^2}}, \quad k = 1, \ldots, n. \tag{5.5}$$

For some user-defined relaxation parameter δ, if $\Delta u_{t_k} < \delta$, then the time spacing is geometrically increased (i. e., $\Delta t \to 2\Delta t$), upto some prefixed value $\Delta t_{\max}$.

5.2.3 *Spatial Approximation*

The gradients and Laplacian in equation (5.2) are spatially approximated using the upwind difference scheme and the second-order central difference scheme, respectively. Recall that the central difference scheme approximation of the convective terms in Eq. (5.2) does not model the flow-physics accurately [37]. Furthermore, a first-order upwind scheme is too diffusive, thereby necessitating the use of higher-order upwind schemes. For example,

$$K_1(\mathbf{x}, t)\frac{\partial u}{\partial x} \approx K_{1ij}\left(\frac{u^k_{i+1,j} - u^k_{i-1,j}}{2\Delta x}\right) + q(K_1^+ u_x^- + K_1^- u_x^+), \tag{5.6}$$

where

$$K_1^- = \min(K_{1ij}, 0) \quad K_1^+ = \max(K_{1ij}, 0),$$
$$u_x^- = \frac{u^k_{i-2,j} - 3u^k_{i-1,j} + 3u^k_{i,j} - u^k_{i+1,j}}{3\Delta x} \quad u_x^+ = \frac{u^k_{i-1,j} - 3u^k_{i,j} + 3u^k_{i+1,j} - u^k_{i+2,j}}{3\Delta x}, \tag{5.7}$$

where the parameter $q = 0.5$ represents the third-order accurate upwind formula (UD3), which is utilized for the interior stencil points. The use of ghost points is avoided by setting $q = 0$ for grid points immediately adjacent to the boundary, leading to a second-order accurate central difference scheme at these points. Next, we introduce a Riemannian metric as well as three scalar invariants associated with this metric to aid the visualization of macrostructures in viscoelastic sub-diffusive flows.

5.3 Time-Asymptotic Stability Analysis and Spectral Analysis

The linear 1D FADR Eq. (5.2) (with $K_1 = c$, $K_2 = \gamma$, $f(x,t) = \lambda u(x,t)$ for $x \in (-\infty, \infty)$, where c, γ and λ are constants specifying the advection speed, coefficient of diffusion and coefficient of reaction, respectively) serves as a model for fractional partial differential equations replicating multi-scale processes.

First, we show that the solution of the linear 1D FADR equation, coupled with periodic boundary conditions, discretized using the numerical method outlined in Sects. 5.2.1–5.2.3, is time-asymptotically stable for a limited range of CFL numbers and Peclet numbers. The discretization of the linear 1D FADR equation, using the θ-method, is as follows,

$$\frac{(\Delta t)^{-\alpha}}{\Gamma(2-\alpha)}\left\{\sum_{j=1}^{n}\left(j^{(1-\alpha)}-(j-1)^{(1-\alpha)}\right)\left(u_i^{n-j+1}-u_i^{n-j}\right)\right\}+c\left(\frac{u_{i+1}^{n-1}-u_{i-1}^{n-1}}{2\Delta x}\right)$$
$$+qc\left(\frac{u_{i-2}^{n-1}-3u_{i-1}^{n-1}+3u_i^{n-1}-u_{i+1}^{n-1}}{3\Delta x}\right)$$
$$=\gamma\theta\frac{u_{i+1}^{n}-2u_i^{n}+u_{i-1}^{n}}{(\Delta x)^2}+\gamma(1-\theta)\frac{u_{i+1}^{n-1}-2u_i^{n-1}+u_{i-1}^{n-1}}{(\Delta x)^2}+\lambda u_i^{n-1}, \tag{5.8}$$

where n/i, denotes the temporal/spatial index, and we set the parameter $q = 0.5$. Introducing the following non-dimensional parameters: the fractional CFL number, N_c, the fractional Peclet number, Pe and the fractional Damköhler number, Da, where

$$N_c = \frac{c(\Delta t)^{\alpha}}{\Delta x}, \quad Pe = \frac{\gamma(\Delta t)^{\alpha}}{(\Delta x)^2}, \quad Da = \frac{\lambda(\Delta x)}{c}. \tag{5.9}$$

In subsequent discussion, we drop the prefix 'fractional' in the above three parameters. Rearranging the terms in Eq. (5.8), we arrive at the following equation,

$$(1+2Pe\theta\Gamma(2-\alpha))\,u_i^n - Pe\theta\Gamma(2-\alpha)u_{i+1}^n - Pe\theta\Gamma(2-\alpha)u_{i-1}^n = (1-qN_c\Gamma(2-\alpha)$$
$$-2Pe(1-\theta)\Gamma(2-\alpha)+DaN_c\Gamma(2-\alpha))\,u_i^{n-1}+\Big(-N_c\frac{\Gamma(2-\alpha)}{2}+qN_c\frac{\Gamma(2-\alpha)}{3}$$
$$+Pe(1-\theta)\Gamma(2-\alpha)\Big)u_{i+1}^{n-1}+\Big(N_c\frac{\Gamma(2-\alpha)}{2}+qN_c\Gamma(2-\alpha)+Pe(1-\theta)\Gamma(2-\alpha)\Big)$$
$$u_{i-1}^{n-1}-qN_c\frac{\Gamma(2-\alpha)}{3}u_{i-2}^{n-1}+\sum_{j=2}^{n}\left(j^{(1-\alpha)}-(j-1)^{(1-\alpha)}\right)\left(u_i^{n-j+1}-u_i^{n-j}\right). \tag{5.10}$$

If $\widetilde{u_i^n}$ be the (numerically) approximate solution of equation (5.10), then the round-off error, $\varepsilon_i^n = u_i^n - \widetilde{u_i^n} \quad (i = 0, \ldots, M)$, identically satisfies the same equation.

$$(1+2Pe\theta\Gamma(2-\alpha))\varepsilon_i^n - Pe\theta\Gamma(2-\alpha)\varepsilon_{i+1}^n - Pe\theta\Gamma(2-\alpha)\varepsilon_{i-1}^n = (1-qN_c\Gamma(2-\alpha)$$
$$-2Pe(1-\theta)\Gamma(2-\alpha)+DaN_c\Gamma(2-\alpha))\varepsilon_i^{n-1} + \left(-N_c\frac{\Gamma(2-\alpha)}{2}+qN_c\frac{\Gamma(2-\alpha)}{3}\right.$$
$$\left.+Pe(1-\theta)\Gamma(2-\alpha)\right)\varepsilon_{i+1}^{n-1} + \left(N_c\frac{\Gamma(2-\alpha)}{2}+qN_c\Gamma(2-\alpha)+Pe(1-\theta)\Gamma(2-\alpha)\right)$$
$$\varepsilon_{i-1}^{n-1} - qN_c\frac{\Gamma(2-\alpha)}{3}\varepsilon_{i-2}^{n-1} - \sum_{j=2}^{n}(j^{(1-\alpha)}-(j-1)^{(1-\alpha)})(\varepsilon_i^{n-j+1}-\varepsilon_i^{n-j}). \quad (5.11)$$

Due to periodic boundary conditions, we have,

$$\varepsilon_0^n = \varepsilon_M^n \quad n = 1, 2, \ldots\ldots, T. \quad (5.12)$$

Assuming the round-off error has the following form,

$$\varepsilon_i^n = \xi_n e^{Ik(ih)}, \quad (5.13)$$

where $\xi_n = |\varepsilon_i^n|$, $I = \sqrt{-1}$, $h = \Delta x$ and $k = \frac{2\pi l}{L}$ (l: index, and L: spatial domain length). We substitute the above relation in the Eq. (5.11) for the round-off error to arrive at the following form,

$$\Rightarrow (1+2Pe\theta\Gamma(2-\alpha))\xi_n e^{Ik(ih)} - Pe\theta\Gamma(2-\alpha)\xi_n e^{Ik(i+1)h} - Pe\theta\Gamma(2-\alpha)\xi_n e^{Ik(i-1)h}$$
$$= \left(1-qN_c\Gamma(2-\alpha)-2Pe(1-\theta)\Gamma(2-\alpha)+DaN_c\Gamma(2-\alpha)\right)\xi_{n-1}e^{Ik(ih)} + \left(-N_c\right.$$
$$\left.\frac{\Gamma(2-\alpha)}{2}+qN_c\frac{\Gamma(2-\alpha)}{3}+Pe(1-\theta)\Gamma(2-\alpha)\right)\xi_{n-1}e^{Ik(i+1)h} + \left(N_c\frac{\Gamma(2-\alpha)}{2}+qN_c\right.$$
$$\left.\Gamma(2-\alpha)+Pe(1-\theta)\Gamma(2-\alpha)\right)\xi_{n-1}e^{Ik(i-1)h} - qN_c\frac{\Gamma(2-\alpha)}{3}\xi_{n-1}e^{Ik(i-2)h} -$$
$$\sum_{j=2}^{n}(j^{(1-\alpha)}-(j-1)^{(1-\alpha)})(\xi_{n-j+1}-\xi_{n-j})e^{Ik(ih)}. \quad (5.14)$$

which can be written in a concise form as follows,

$$\mu_1\xi_n = \mu_2\xi_{n-1} + \sum_{j=2}^{n} r_j^{\alpha}\left(\xi_{n-j+1}-\xi_{n-j}\right), \quad (5.15)$$

where the coefficients are given by,

$$\mu_1 = 1 + 2Pe\theta\Gamma(2-\alpha)(1-\cos(kh)),$$
$$\mu_2 = 1 - \Bigg(N_c q \left(1 - \frac{2}{3}\cos(kh)\right) + 2Pe(1-\theta)(1-\cos(kh)) - DaN_c + IN_c\sin(kh)$$
$$- \frac{N_c q}{3}\left(2e^{-Ikh} - e^{-2Ikh}\right)\Bigg)\Gamma(2-\alpha),$$
$$r_j^{\alpha} = j^{(1-\alpha)} - (j-1)^{(1-\alpha)}. \tag{5.16}$$

Next, we prove the result that the finite difference scheme (5.15–5.16) is time-asymptotically stable, in the form of the following theorem.

Theorem 5.1 *The approximate solution to the 1D linear FADR equation using the θ-method* (5.15) *with $\alpha \in (0, 1)$, on the finite domain $x \in [-L, L]$ with periodic boundary conditions is time-asymptotically stable for all $t \geq 0$.*

Proof It suffices for us to show that the time-amplification factor, ξ_n (Eq. (5.15)), obeys the inequality,

$$\xi_n \leq \xi_{n-1} \leq \xi_{n-2} \leq \ldots \leq \xi_1 \leq \xi_0. \tag{5.17}$$

- For $n = 1$ in Eqs. (5.15, 5.16), we observe the $\mu_1 \geq 1$ while $\mu_2 \leq 1$ for sufficiently small N_c, and Pe, thereby leading us towards the conclusion that $\xi_1 \leq \xi_0$.
- Using induction hypothesis, we assume that $\xi_{n-1} \leq \xi_{n-2} \leq \ldots \leq \xi_1 \leq \xi_0$. Next we show that $\xi_n \leq \xi_{n-1}$.
- Since $\mu_1 \geq 1$ and $\mu_2 \leq 1$, Eq. (5.15) can be replaced with the following inequality,

$$\xi_n \leq \xi_{n-1} + \sum_{j=2}^{n} r_j^{\alpha}\left(\xi_{n-j+1} - \xi_{n-j}\right) \leq \xi_{n-1}. \tag{5.18}$$

The second inequality in equation (5.18) is because,

1. r_j^{α} is positive and
2. $r_j^{\alpha} > r_{j+1}^{\alpha}$, and
3. $\xi_{n-j+1} - \xi_{n-j} \leq 0$ (via the induction hypothesis),

which completes the proof. □

We emphasize that while the θ-method is not unconditionally stable, it is time-asymptotically stable for a restricted range of N_c and Pe.

Spectral Analysis

Although the θ-method is time-asymptotically stable, the presence of dispersion errors (through negative group velocity and large phase speed errors) would invalidate the long-time integration results [39]. Hence, we couple the result in Sect. 5.3 along with the toolset of spectral analysis to deduce the relevant range of the parameters,

N_c, Pe and Da, for an accurate representation of the numerical solution of the 1D FADR equation.

Using the spectral (Fourier) representation of the approximate solution of equation (5.10), we have $\widetilde{u_i^n} = \xi'_n e^{I(ikh)}$ $(I = \sqrt{-1},\ k = \frac{2\pi I}{L})$. We define the numerical time-amplification factor, $G_{\text{num}} = \frac{\widetilde{u_i^n}}{\widetilde{u_i^{n-1}}} = \frac{\xi'_n}{\xi'_{n-1}}$. Dividing equation (5.10) with $\widetilde{u_i^{n-1}}$, we arrive at the equation governing G_{num},

$$\begin{aligned}&\Big(1 + 2Pe\theta\Gamma(2-\alpha)(1-\cos(kh))\Big)G_{\text{num}}=\Big(1-qN_c\Gamma(2-\alpha)-2Pe(1-\theta)\Gamma(2-\alpha)\\&+ DaN_c\Gamma(2-\alpha)\Big)-IN_c\Gamma(2-\alpha)\sin(kh)+\frac{2qN_c\Gamma(2-\alpha)}{3}\cos(kh)+\frac{2}{3}qN_c\Gamma(2-\alpha)\\&e^{-Ikh}+2Pe(1-\theta)\Gamma(2-\alpha)\cos(kh)-\frac{qN_c}{3}\Gamma(2-\alpha)e^{-2Ikh}\\&-\sum_{j=2}^{n}\left(j^{(1-\alpha)}-(j-1)^{(1-\alpha)}\right)\left(\frac{1}{G_{\text{num}}^{j-2}}-\frac{1}{G_{\text{num}}^{j-1}}\right).\end{aligned}\tag{5.19}$$

$$\begin{aligned}&(1+2Pe\theta\Gamma(2-\alpha)(1-\cos(kh)))\,G^n_{\text{num}} = \Bigg[1 - qN_c\Gamma(2-\alpha)\left(1-\frac{4}{3}\cos(kh)\right.\\&\left.+\frac{1}{3}\cos(2kh)\right) - 2Pe(1-\theta)\Gamma(2-\alpha)(1-\cos(kh)) + DaN_c\Gamma(2-\alpha)\\&- IN_c\Gamma(2-\alpha)\left(\sin(kh)+\frac{2}{3}q\sin(kh)-\frac{q}{3}\sin(2kh)\right)\Bigg]G^{n-1}_{\text{num}}\\&-\sum_{j=2}^{n}(j^{(1-\alpha)}-(j-1)^{(1-\alpha)})\left(G^{n-j+1}_{\text{num}}-G^{n-j}_{\text{num}}\right).\end{aligned}\tag{5.20}$$

Thus, we have,

$$C_0G^n_{\text{num}} + C_1G^{n-1}_{\text{num}} + \sum_{j=2}^{n} r_j^{\alpha}\left(G^{n-j+1}_{\text{num}} - G^{n-j}_{\text{num}}\right) = 0,\tag{5.21}$$

where the coefficients,

$$\begin{aligned}C_0 =&1+2Pe\theta\Gamma(2-\alpha)(1-\cos(kh))\\C_1=&-1+qN_c\Gamma(2-\alpha)\left(1-\frac{4}{3}\cos(kh)+\frac{1}{3}\cos(2kh)\right)+2Pe(1-\theta)\Gamma(2-\alpha)\\&(1-\cos(kh))-DaN_c\Gamma(2-\alpha)+IN_c\Gamma(2-\alpha)\left(\sin(kh)+\frac{2}{3}q\sin(kh)-\frac{q}{3}\sin(2kh)\right).\end{aligned}\tag{5.22}$$

We remark that the order 'n' of the polynomial (5.21) is fixed at $n = 75$ in the subsequent analysis since our numerical studies have shown that an increase in the polynomial order by one shifts the contours of the spectral variables by less than 0.001%.

Next, transforming the exact solution of the linear 1D FADR equation, using Fourier-Laplace transform as $u(x,t) = \int\int \hat{U}(k,\omega)e^{I(kx-\omega t)}$, we arrive at the exact dispersion relation,

$$\omega_{\text{exact}} = \omega = I(\lambda - \gamma k^2 - cIk)^{\frac{1}{\alpha}}. \tag{5.23}$$

Similarly, we have a corresponding numerical dispersion relation for the approximate equation (5.10),

$$\omega_{\text{num}} = I(\lambda^N - \gamma^N k^2 - c^N Ik)^{\frac{1}{\alpha}}, \tag{5.24}$$

where the superscript 'N' denotes the corresponding numerical values of the parameters. Since

$$G_{\text{num}} = \frac{u(k,t+\Delta t)}{u(k,t)} = \frac{e^{I(kx-\omega_{\text{num}}(t+\Delta t))}}{e^{I(kx-\omega_{\text{num}}t)}} = e^{-I\omega_{\text{num}}\Delta t} = e^{I\beta}, \tag{5.25}$$

and the numerical phase speed is given by, $c_{num} = \frac{\omega_{num}}{k}$, we find that,

$$c_{\text{num}} = -\frac{\beta}{k\Delta t} = -\frac{1}{k\Delta t}\tan^{-1}\left(\frac{(G_{\text{num}})_i}{(G_{\text{num}})_r}\right), \tag{5.26}$$

where the subscript 'Imag'/'Re' denotes the imaginary/real values, respectively. Using the expression for the exact phase speed, $c_{exact} = \frac{\omega_{exact}}{k}$ and the non-dimensional parameters described in equation (5.10), we find the ratio of the phase speeds,

$$\frac{c_{\text{num}}}{c_{\text{exact}}} = \frac{-I\beta}{(DaN_c - (kh)^2 Pe - I(kh)N_c)^{\frac{1}{\alpha}}}, \tag{5.27}$$

Finally, the expression for the numerical and the exact group velocities are given by,

$$\begin{aligned} V_{g\,\text{num}} &= \left[\frac{\partial}{\partial k}\omega_{\text{num}}\right]_r = \left[\frac{\partial}{\partial k}\left(\frac{\beta}{\Delta t}\right)\right]_r = \left[\frac{h}{\Delta t}\frac{d\beta}{d(kh)}\right]_r, \\ V_{g\,\text{exact}} &= \left[\frac{\partial}{\partial k}\omega_{\text{exact}}\right]_r = \left[\frac{\partial}{\partial k}\left(I(\lambda - \gamma k^2 - cIk)^{\frac{1}{\alpha}}\right)\right]_r = \left[\frac{1}{\alpha}(c - 2k\gamma I)(-I\omega)^{1-\alpha}\right]_r. \end{aligned} \tag{5.28}$$

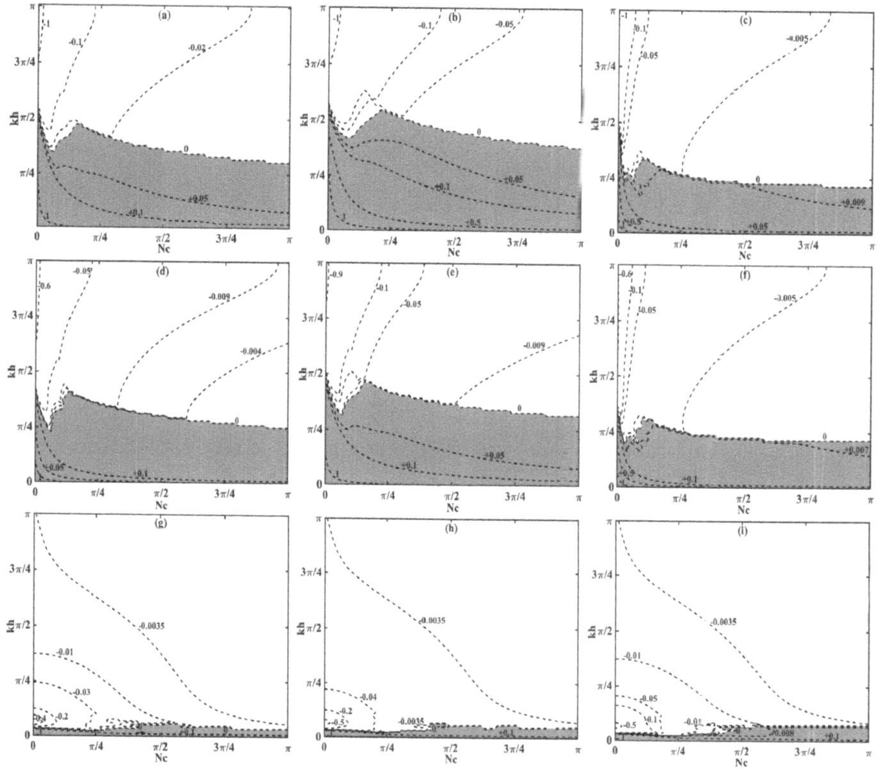

Fig. 5.1 Viscoelastic sub-diffusive stability phase diagram in the $Re - We$ parametric space for the Rouse model (solid curves) and for the Zimm's model (dashed curves), evaluated at transverse spatial locations: **a**, **b** $y = 0.2$, **c**, **d** $y = 0.5$, **e**, **f** $y = 0.7$ and **g**, **h** $y = 0.9$ and at fixed values of viscosity ratios, $\nu = 0.05$ (left column) and $\nu = 0.3$ (right column)

Hence,

$$\frac{(V_g)_{\text{num}}}{(V_g)_{\text{exact}}} = \left(\frac{\alpha}{\begin{array}{l} N_c \left(r^{1-\alpha} (\Delta t)^{1-\alpha} \cos{(1-\alpha)\phi} \right) \\ \quad + 2kh \left(r^{1-\alpha} (\Delta t)^{1-\alpha} \sin{(1-\alpha)\phi} \right) Pe \end{array}} \right) \left(\frac{d\beta}{d(kh)} \right) \Big|_r , \quad (5.29)$$

where $\omega = r e^{I(\phi + \frac{\pi}{2})}$.

The contour plots for the group velocity ratio, $V_g = \left[\frac{V_{g,\text{num}}}{V_{g,\text{exact}}} \right]_{Re}$, for Peclet numbers, $Pe = 0.001, 0.01$ and 1.0, and Damköhler numbers, $Da = -0.01, 0.0$ and 0.01, are presented for two values of θ, namely, $\theta = 0.5$ (in Fig. 5.1) and $\theta = 1.0$ (in Fig. 5.2) in the (N_c, kh)-plane. The corresponding contour plots for the absolute phase speed error, $\Delta c = |1 - \frac{c_{\text{num}}}{c_{\text{exact}}}|$ are shown in Figs. 5.3 and 5.4 for $\theta = 0.5$

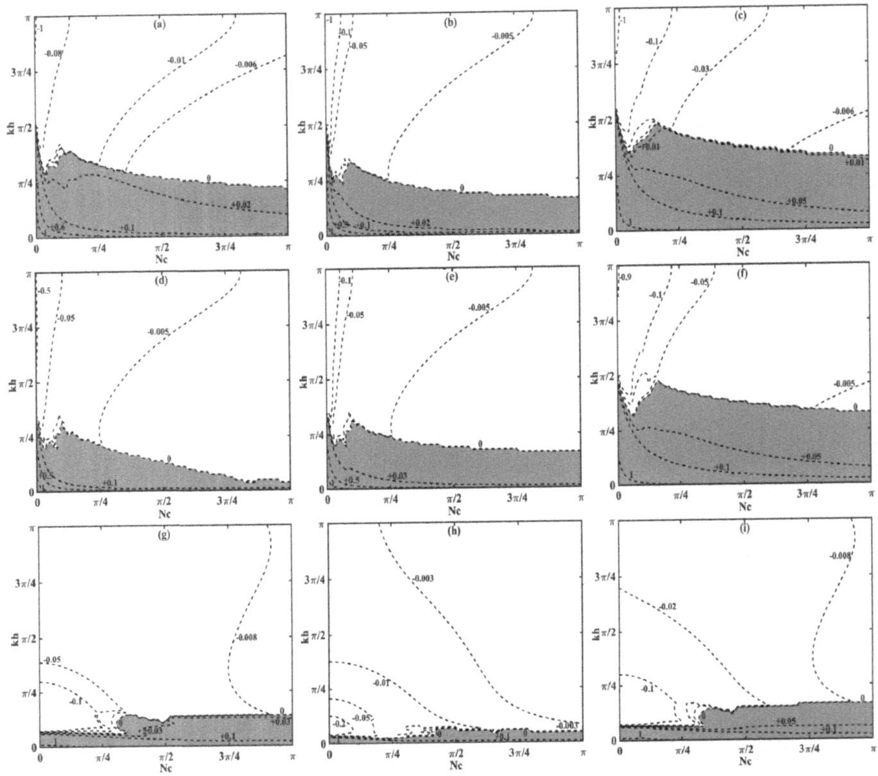

Fig. 5.2 Group velocity ratio contours, V_g, at $\theta = 1.0$, $\alpha = 0.9$ and **a** $Pe = 0.001$, $Da = -0.01$, **b** $Pe = 0.001$, $Da = 0.0$, **c** $Pe = 0.001$, $Da = 0.01$, **d** $Pe = 0.01$, $Da = -0.01$, **e** $Pe = 0.01$, $Da = 0.0$, **f** $Pe = 0.01$, $Da = 0.01$, **g** $Pe = 1.0$, $Da = -0.01$, **h** $Pe = 1.0$, $Da = 0.0$ and **i** $Pe = 1.0$, $Da = 0.01$

and $\theta = 1.0$, respectively. Regions of favourable spectral properties ($V_g > 0$ and $\Delta c \leq 0.1$) are highlighted with grey colour.

We outline the spectral properties, namely, the group velocity ratio and the absolute phase speed error, for a specific case of the fractional order, $\alpha = 0.9$. In general, we find that in the limit of vanishingly small values of kh, the numerical method has favourable spectral properties (i. e., $V_g > 0$ and $\Delta c \leq 0.1$ in the limit, $kh \to 0$). The region of positive group velocities ($V_g > 0$) indicates a region where the numerical solution travels in the correct direction and the numerical instabilities in the form of q-waves are avoided [37]. Figures 5.1 and 5.2 indicate that this favourable region is restricted to smaller values of kh with larger values of Pe as well as with smaller values of θ. While the former observation can be attributed to the fact that a stiff diffusive term (i. e., larger '$K_2(\mathbf{x}, t)\nabla^2 u(\mathbf{x}, t)$' term in equation (5.2) can be correctly approximated with smaller grid-size, h; the latter observation can be explained through a numerical stabilization due to the implicit treatment of fast

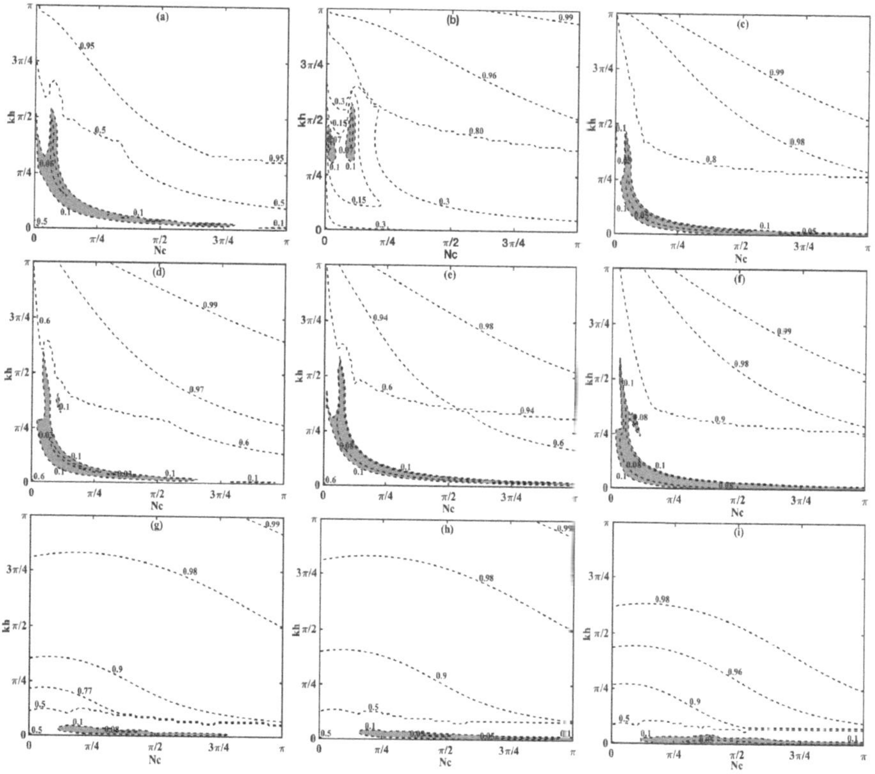

Fig. 5.3 Absolute phase speed error contours, Δc at $\theta = 0.5$, $\alpha = 0.9$ and **a** $Pe = 0.001$, $Da = -0.01$, **b** $Pe = 0.001$, $Da = 0.0$, **c** $Pe = 0.001$, $Da = 0.01$, **d** $Pe = 0.01$, $Da = -0.01$, **e** $Pe = 0.01$, $Da = 0.0$, **f** $Pe = 0.01$, $Da = 0.01$, **g** $Pe = 1.0$, $Da = -0.01$, **h** $Pe = 1.0$, $Da = 0.0$ and **i** $Pe = 1.0$, $Da = 0.01$

time scale term (in this case, the diffusive term). Both of these observations are in congruence with the corresponding analysis of the integer order ADR equations [37]. Similarly, the absolute phase error contours highlight spectrally favourable region ($\Delta c \leq 0.1$, equivalently the phase errors are restricted to 10% of the exact phase speed, see Figs. 5.3 and 5.4) at lower values of Pe and at lower values of θ, indicating the fact that the phase errors in this numerical method can be reduced by introducing finer grids [40, 41].

We summarize our discussion by indicating two sources of error that are particularly perplexing in the direct numerical simulations of FADR equations. The first source of error is the existence of q-waves for those sets of numerical parameters for which the spatiotemporal discretization is time-asymptotically stable. In such a situation, the q-waves do not attenuate and have to be eliminated by deploying an explicit filter [42]. Another aspect of spectral error is related to the Gibbs' phenomenon, which occurs as a consequence of sharp discontinuity in the solution and which

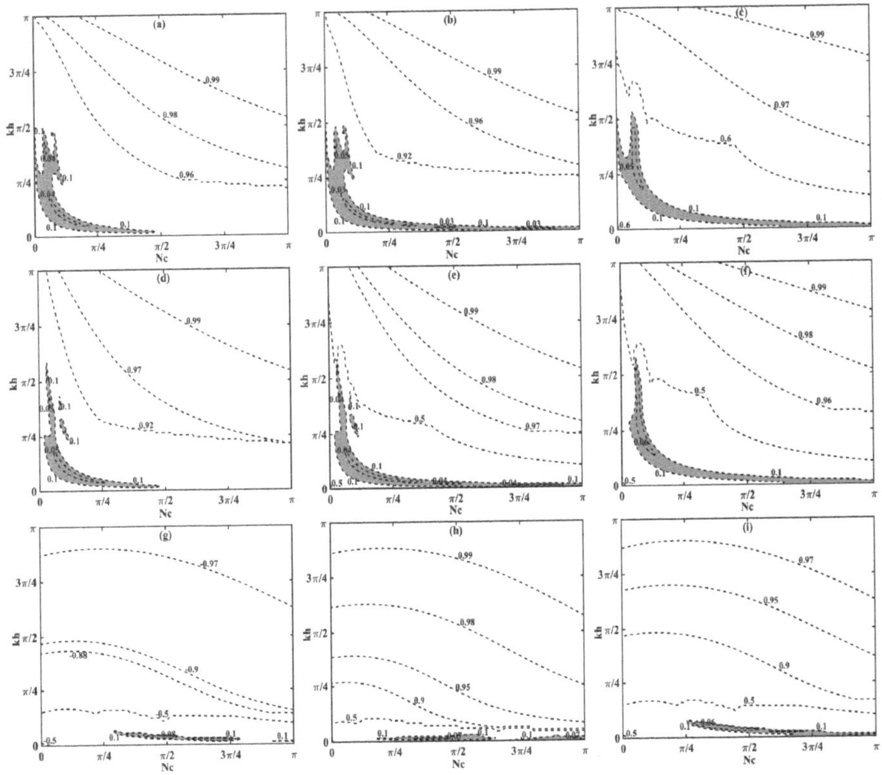

Fig. 5.4 Absolute phase speed error contours, Δc at $\theta = 1.0$, $\alpha = 0.9$ and **a** $Pe = 0.001$, $Da = -0.01$, **b** $Pe = 0.001$, $Da = 0.0$, **c** $Pe = 0.001$, $Da = 0.01$, **d** $Pe = 0.01$, $Da = -0.01$, **e** $Pe = 0.01$, $Da = 0.0$, **f** $Pe = 0.01$, $Da = 0.01$, **g** $Pe = 1.0$, $Da = -0.01$, **h** $Pe = 1.0$, $Da = 0.0$ and **i** $Pe = 1.0$, $Da = 0.01$

causes fictitious oscillations, a problem that can be remedied using high-accuracy dispersion relation preserving schemes (which has at least shown promise in integer order partial differential equations) [39].

5.4 Method Validation: 2D Fractional Diffusion Equation

Using the spectrally relevant parameter values discussed in Sect. 5.3, the θ-method is verified by solving the 2D fractional diffusion equation ($K_1 = f = 0$ and $K_2 = 1$ in equation (5.2)), originally proposed by Brunner [26], for the case of (a) zero Dirichlet boundary conditions ($\alpha_1 = 1, \alpha_2 = 0, g_0(\mathbf{x}, t) = 0$) and (b) zero Neumann conditions ($\alpha_1 = 0, \alpha_2 = 1, g_0(\mathbf{x}, t) = 0$), over the square domain, $(x, y) \in \Gamma = [-1, 1]^2$. The initial condition for the two test cases can be outlined as follows,

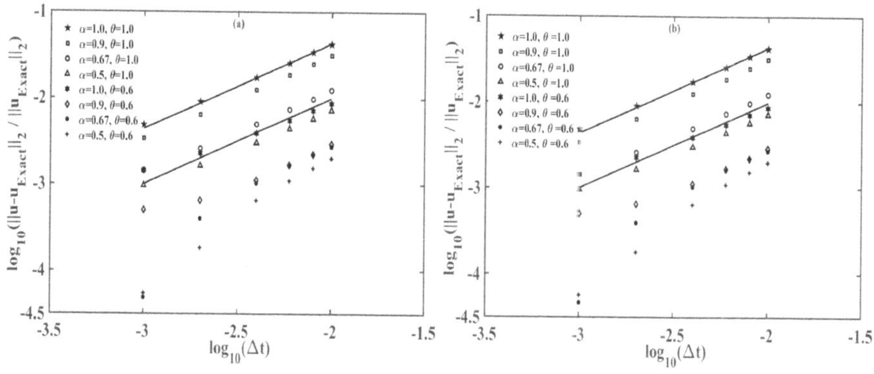

Fig. 5.5 Relative error for the solution of the 2D fractional diffusion equation at simulation time $T = 0.35$, at $\alpha = 1.0$, $\theta = 1.0$ ($\star$); $\alpha = 0.9$, $\theta = 1.0$ ($\square$); $\alpha = 0.67$, $\theta = 1.0$ ($\circ$); $\alpha = 0.5$, $\theta = 1.0$ ($\triangle$); $\alpha = 1.0$, $\theta = 0.6$ ($*$); $\alpha = 0.9$, $\theta = 0.6$ ($\diamond$); $\alpha = 0.67$, $\theta = 0.6$ ($\bullet$); $\alpha = 0.5$, $\theta = 0.6$ ($+$); and for **a** Dirichlet boundary conditions, and **b** Neumann boundary conditions

1. **Dirichlet case:** $u_0(x, y) = \cos\left(\frac{\pi}{2}x\right)\cos\left(\frac{\pi}{2}y\right)$,
2. **Neumann case:** $u_0(x, y) = \sin\left(\frac{\pi}{2}x\right)\sin\left(\frac{\pi}{2}y\right)$,

and the exact solution for both cases is given by

$$u_{\text{Exact}}(x, y, t) = E_\alpha\left(-\frac{1}{4}\pi^2 t^\alpha\right) u_0(x, y), \tag{5.30}$$

where $E_\alpha(z) = \sum_{k=0}^{\infty} \frac{z^k}{\Gamma(\alpha k+1)}$, $\alpha > 0$, is the one-parameter Mittag Leffler function.

The domain, $\Gamma \cup \partial\Gamma$, is discretized using 51×51 points and the relative error in l_2 – norm, $\frac{\|\tilde{u}-u_{\text{Exact}}\|_2}{\|u_{\text{Exact}}\|_2}$, versus discretization time, Δt, is shown in Fig. 5.5. The total computation time is fixed at $T = 0.35$ for all cases. We note that the slope of the error curves at $\alpha = 1.0$ is one (highlighted with solid curves in Fig. 5.5a, b). This observation can be explained since at this value of α, the θ-method reduces to the standard, integer order, Euler method, which is first-order accurate. Further, note that the slope of the error curves for $\alpha < 1.0$ is sub-linear, indicating that for the fractional-order diffusion equation, the θ-method has a sub-linear order of accuracy.

5.5 Direct Numerical Simulation: Planar Fractional Viscoelastic Channel Flow

Next, we describe the incompressible, sub-diffusive dynamics of a planar viscoelastic channel flow for polymer melts and present the numerical solution using the θ-method described in Sect. 5.2. Using the following scales for non-dimensionalizing the governing equations: the height of the channel H for length, the timescale T corresponding to maximum base flow velocity (i. e., $T = (H/\mathcal{U}_0)^{1/\alpha})$ for time and $\rho\mathcal{U}_0^2$ for pressure and stresses (where $\rho, \mathcal{U}_0$ is the density and the velocity scale, respectively), we summarize the model in streamfunction-vorticity formulation as follows,

$$Re\left[\frac{\partial^\alpha \Omega}{\partial t^\alpha} + \boldsymbol{v}\cdot\nabla\Omega\right] = \nu\nabla^2\Omega + \frac{(1-\nu)}{We}\nabla\times\nabla\cdot\mathcal{C}, \tag{5.31a}$$

$$\nabla^2\psi = -\Omega, \tag{5.31b}$$

$$\frac{\partial^\alpha \mathcal{C}}{\partial t^\alpha} + \boldsymbol{v}\cdot\nabla\mathcal{C} - (\nabla\boldsymbol{v})^T\mathcal{C} - \mathcal{C}\nabla\boldsymbol{v} = \frac{\boldsymbol{I}-\mathcal{C}}{We}, \tag{5.31c}$$

where the variables $t, \Psi, \mathbf{v} = (u, v) = (\frac{\partial\Psi}{\partial y}, -\frac{\partial\Psi}{\partial x}), \mathcal{C}, \Omega = \nabla\times\mathbf{v}$ denote time, streamfunction, velocity, polymer conformation tensor and vorticity, respectively. The parameters $\eta_s, \eta_p, \eta_0 = \eta_s + \eta_p, \nu = \eta_s/\eta_0$ and μ are the solvent viscosity, the polymeric contribution to the shear viscosity, the total viscosity, the viscous contribution to the total viscosity of the fluid and strength of the body force, respectively. The dimensionless groups characterizing inertia and elasticity are Reynolds number, $Re = \frac{\rho\mathcal{U}_0 H}{\eta_0}$, and Weissenberg number, $We = \frac{\lambda^\alpha\mathcal{U}_0}{H}$, respectively. The parameter, λ^α, is the polymer relaxation time. Note that Eq. (5.31c) represents the fractional version of the regular Oldroyd-B model for viscoelastic fluids [43], derived in a recently reported work [33]. Two specific cases of the fractional derivative are considered, namely, monomer diffusion in Rouse chain melts ($\alpha = 1/2$) [16], and in Zimm chain solution ($\alpha = 2/3$) [44].

5.5.1 *Initial and Boundary Conditions*

Initial conditions: A rectilinear coordinate system is used with x, y denoting the channel flow direction and the transverse direction, respectively. The origin of this coordinate system is chosen at the left end of the lower wall of the channel. The mean flow is assumed to be a plane Poiseuille flow with its variation entirely in the transverse direction, namely,

$$\mathbf{U}_0 = (y - y^2)\mathbf{e_x}, \tag{5.32}$$

where $\mathbf{e_x}$ is the unit vector along x-direction. The mean flow, $\mathbf{U}_0$, defines the mean vorticity, $\Omega_0 = 2y - 1$, and the mean stream-function, $\psi_0 = \frac{y^2}{2} - \frac{y^3}{3}$. The base state polymer conformation tensor, $\mathcal{C}_0 = [\mathcal{C}_{0ij}]$ is,

$$\begin{aligned} \mathcal{C}_{011} &= 1, \\ \mathcal{C}_{012} = \mathcal{C}_{021} &= We(1-2y), \\ \mathcal{C}_{022} &= 2We^2(1-2y)^2 + 1. \end{aligned} \tag{5.33}$$

The initial condition is composed of the mean flow superposed with a perturbed unstable mode, as follows,

$$
\begin{aligned}
\Omega|_{t=0} &= \Omega_0 + \epsilon\Omega_1 = \Omega_0 + \epsilon\{\widetilde{\Omega}(y)|_{t=0}e^{ikx}\}_r,\\
\psi|_{t=0} &= \psi_0 + \epsilon\psi_1 = \psi_0 + \epsilon\{\widetilde{\psi}(y)|_{t=0}e^{ikx}\}_r,\\
\mathcal{C}|_{t=0} &= \mathcal{C}_0 + \epsilon\mathcal{C}_1 = \mathcal{C}_0 + \epsilon\{\widetilde{\mathcal{C}_1}(y)|_{t=0}e^{ikx}\}_r,
\end{aligned}
\tag{5.34}
$$

where $(\Omega_1, \psi_1, \mathcal{C}_1)$ are the perturbations that are Fourier transformed in the x-direction. The subscript, r, denotes the real part of the complex valued function. The equations governing the initial conditions (5.34) can be obtained by assuming a normal mode expansion for the perturbed field, $\phi_1 = \widetilde{\phi(y)}e^{ikx}$ (where $\phi_1 = (\Omega_1, \psi_1, \mathcal{C}_1)$, Eq. (5.31) in $\mathcal{O}(\epsilon)$ reduces to,

$$
\begin{aligned}
&Re\left((y-y^2)(ik)\widetilde{\Omega} - 2(ik)\widetilde{\psi}\right) - \nu\left((ik)^2\widetilde{\Omega} + \widetilde{\Omega}''\right)\\
&\quad - (1-\nu)\left((ik)^2\widetilde{\mathcal{C}}_{12} - \widetilde{\mathcal{C}}''_{12} + (ik)\widetilde{\mathcal{C}}'_{22} - (ik)\widetilde{\mathcal{C}}'_{11}\right) = 0, && (5.35a)\\
&\widetilde{\psi}(ik)^2 + \widetilde{\psi}'' + \widetilde{\Omega} = 0, && (5.35b)\\
&(y-y^2)(ik)\widetilde{\mathcal{C}}_{11} - 2(ik)\widetilde{\psi}' + 2We(1-2y)(ik)^2\widetilde{\psi}) + \frac{\widetilde{\mathcal{C}}_{11}}{We} = 0, && (5.35c)\\
&(y-y^2)(ik)\widetilde{\mathcal{C}}_{12} + 2We(ik)\widetilde{\psi} + (2We^2(1-2y)^2 + 1)(ik)^2\widetilde{\psi} - (1-2y)\widetilde{\mathcal{C}}_{11}\\
&\quad - \widetilde{\psi}'' + \frac{\widetilde{\mathcal{C}}_{12}}{We} = 0, && (5.35d)\\
&(y-y^2)(ik)\widetilde{\mathcal{C}}_{22} + 8We^2(1-2y)(ik)\widetilde{\psi} - 2(1-2y)\widetilde{\mathcal{C}}_{12} - 2We(1-2y)\widetilde{\psi}''\\
&\quad + 2(2We^2(1-2y)^2 + 1)(ik)\widetilde{\psi}' + \frac{\widetilde{\mathcal{C}}_{22}}{We} = 0. && (5.35e)
\end{aligned}
$$

where we denote $\frac{d}{dy}() = ()'$. The solution to the boundary value problem is found subject to the boundary conditions, $(\widetilde{\psi}(y), \widetilde{\psi}'(y)) = (0, 0)$ at the rigid walls $y = 0, 1$, using a standard MATLAB boundary value solver.

Boundary conditions: In order to imitate an infinitely long channel, periodic boundary conditions are assumed at the flow inlet and outlet. No-slip (i. e., $u = v = 0$) and zero tangential conditions (i. e., $\frac{\partial u}{\partial x} = \frac{\partial v}{\partial x} = 0$) are imposed on the lower wall ($y = 0$) and the upper wall ($y = 1.0$) of the channel, respectively. Further, incompressibility constraint provides an additional condition on the walls: $\frac{\partial v}{\partial y} = 0$. Since the flow is parallel to the channel walls, the walls may be treated as streamline. Thus, the streamfunction value, Ψ, on the wall is set as a constant. That constant (which may be different on the lower and the upper wall) is found from the no-slip condition. Zero tangential conditions imply that all tangential derivatives of streamfunction vanish on the wall. Thus, the boundary condition for vorticity is found from the Poisson equation (5.31b),

$$
\frac{\partial^2\Psi}{\partial y^2}|_{\text{wall}} = -\Omega_{\text{wall}}. \tag{5.36}
$$

Finally, the boundary conditions for the elastic stress tensor are constructed from Eq. (5.31c), coupled with the no-slip and zero tangential conditions, as follows,

$$
\begin{aligned}
&\mathcal{C}_{11} = 1, \\
&\frac{\mathcal{C}_{12}}{We} + \frac{\partial^{\alpha}\mathcal{C}_{12}}{\partial t^{\alpha}} - \frac{\partial u}{\partial y} = 0, \\
&\frac{\mathcal{C}_{22} - 1}{We} + \frac{\partial^{\alpha}\mathcal{C}_{22}}{\partial t^{\alpha}} - 2\frac{\partial u}{\partial y}\mathcal{C}_{12} = 0.
\end{aligned} \tag{5.37}
$$

5.5.2 Algorithmic Details

The size of the domain is chosen to be $(x, y) \in [0, 5] \times [0, 1]$. The domain is discretized using 76×51 points, such that the discrete points are equally spaced at $\Delta x = \frac{5}{75}$ and $\Delta y = \frac{1}{50}$ (excluding the boundary points, where the Dirichlet boundary conditions are imposed) in the x- and the y- directions, respectively. The variable in the θ-method is fixed at $\theta = 1.0$. The minimum and the maximum time-step are chosen as $\Delta t_{\min} = 10^{-3}$ and $\Delta t_{\max} = 1.6 \times 10^{-2}$, respectively. The parameters in the initial conditions (5.34) are set at $\epsilon = 0.1, k = 1$. The time-step in the simulation is increased adaptively using the technique outlined in Sect. 5.2.2. The Poisson equation (5.31b) is iteratively solved using the Gauss-Siedal iteration technique. Assuming $(N + 1)$ (or $(M + 1)$) points in the x (or y) direction, the size of the coefficient matrix is $N(M - 1) \times N(M - 1)$. Hence, due to size constraints, the numerical inversion of this coefficient matrix imposes severe memory restrictions. Instead, we note that the coefficient matrix has the following block diagonal structure with $(M - 1)$ blocks,

$$
\begin{pmatrix}
D_1 & D_2 & 0 & \cdots & 0 \\
D_2 & \ddots & \ddots & & \vdots \\
0 & \ddots & \ddots & \ddots & 0 \\
\vdots & & \ddots & \ddots & D_2 \\
0 & \cdots & 0 & D_2 & D_1
\end{pmatrix}, \tag{5.38}
$$

where D_1 (D_2) are $N \times N$ tridiagonal (diagonal) matrices, such that $D_1 = \text{Tridiag}(r_1, R, r_1)$ and $D_2 = \text{Diag}\ (r_2)$ and $r_1 = (\Delta x)^{-2}, r_2 = (\Delta y)^{-2}$ and $R = -2(r_1 + r_2)$. Since the exact location of the non-zero entries is known, the invocation of the entire coefficient matrix in the Guass-Siedal iteration is avoided. The solution is updated by utilizing the non-zero entries of each row on a case-by-case basis. The row diagonal dominance of the matrix (5.38) ensures that the iteration converges in a finite number of steps. Similarly, the vorticity equation (5.31a) involves another Laplacian term and its discretization involves a coefficient matrix identical to the form (5.38) (with $r_1 = -\nu\theta(dx)^{-2}$, $r_2 = -\nu\theta(dy)^{-2}$, $R = -2(r_1 + r_2) + \frac{Re(\Delta t)^{-\alpha}}{\Gamma(2-\alpha)}$). Thus, the Laplacian term in equation (5.31a) is treated identically as above.

5.5.3 Numerical Evolution of the Conformation Tensor Equations

The advection term in the conformation tensor equation (5.31c) is second-order accurate in space with the exception of a few points where it is only first-order accurate. This change in order is due to the special treatment of the advection term in the conformation tensor that is needed to avoid the Hadamard instabilities, resulting in the loss of the symmetric, positive definite (SPD) property of the conformation tensor [45]. The special treatment is an efficient transformation of the slope-limiting approach of Vaidyanathan et al. [46], which requires the evaluation of three schemes: forward, backward, and centred stencils, to approximate the advective flux at each grid point and at each time-step. The resultant scheme that maximizes the eigenvalues of the conformation tensor at the grid points is then chosen. Further, following Dubief et al. [47], we ensure that the polymer does not exceed its maximum extensibility by using a semi-implicit approach for the relaxation term in Eq. (5.31c).

5.5.4 Numerical Results

Next, the in silico studies for two specific cases of monomer diffusion in Rouse chain melts ($\alpha = 1/2$) [16], and in Zimm chain solution ($\alpha = 2/3$) [44] are reported. The instantaneous principle invariant of the conformation tensor, tr$\mathcal{C}$ as well as the new invariants (refer previous chapter on the construction of these invariants) for a specific set of parameters values, $\nu = 0.3$, $Re = 70$, $We = 20$, and simulation time, $T = 7.15$, are presented for both the Zimm's case (left column) and the Rouse case (right column) in Fig. 5.6. Figure 5.6c (as well as 5.6d) shows the logarithmic volume ratio, δ_1. This quantity is qualitatively similar to the principle invariant, tr$\mathcal{C}$, and hence we have a visual resemblance in Fig. 5.6a, c. However, we find that in Fig. 5.6c, we have predominantly negative values, indicating that the instantaneous volume is smaller than the volume of the mean conformation. Further, observe regions of very high values of δ_1 interspersed with regions of very low values, especially near the wall. This observation is the result of the slow diffusion of polymers in sub-diffusive flows since there is no direct mechanism for smoothening out these 'elastic shocks' in the tensor field. The measure, δ_1 does not distinguish between volume-preserving deformations. For example, δ_1 does not distinguish between $\mathcal{C}$ and det$(\mathcal{C})\mathcal{C}_1$, for any tensor $\mathcal{C}_1$ with a unit determinant. In particular, $\delta_1 = 0$, does not imply $\mathcal{C} = \overline{\mathcal{C}}$. In order to identify regions where the instantaneous polymer conformation equals the mean conformation and quantify the deviation when it is not, we use the squared

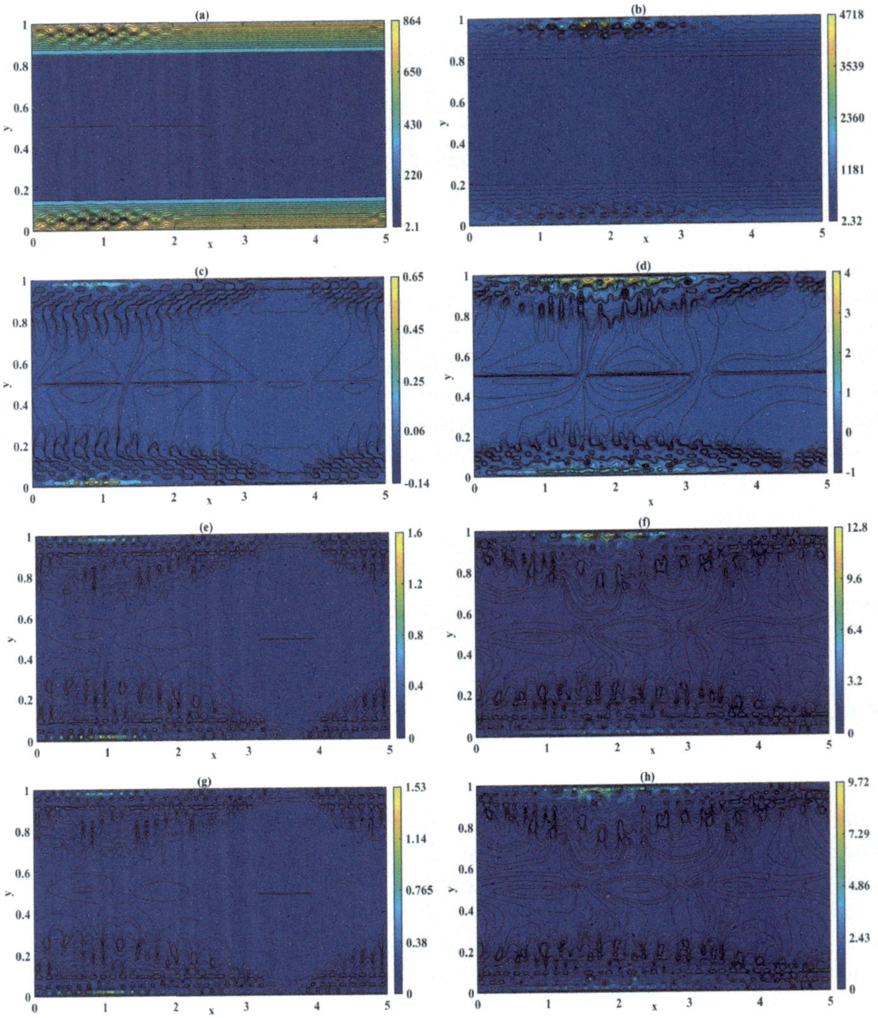

Fig. 5.6 Contours of instantaneous (**a**, **b**) principle invariant of $\mathcal{C}$, tr$\mathcal{C}$, **c**, **d** volume ratio, δ_1, **e**, **f** shortest distance from the mean, δ_2, **g**, **h** anisotropy index, δ_3, for the Zimm's model (left column) and the Rouse model (right column) at simulation time, $T = 7.15$. Other parameters are fixed at $\nu = 0.3. Re = 70$, $We = 20$

geodesic distance away from the mean along the Riemannian manifold, δ_2 (Fig. 5.6e). Figure 5.6e indicates that the conformation tensor field, C, is significantly far away from $\overline{C}$, near the wall. This deviation of δ_2, in the near wall region, can be explained via the 'memory effect', previously observed in regular Oldroyd-B fluids [43]. Finally, Fig. 5.6g shows the instantaneous contours of the anisotropy index, δ_3. This index shows how close the shape of the instantaneous conformation tensor is to the shape of the mean conformation tensor, irrespective of volumetric changes. The visual resemblance of δ_2 and δ_3 suggests that deformations to the mean conformation are largely anisotropic, near the wall. Next, we limit our focus on the contours of the first invariant, δ_1 (after noting in Fig. 5.6 that the contours of the other invariants are qualitatively similar) and for two different values of ν: $\nu = 0.3$ (elastic stress dominated case) and $\nu = 0.5$ (viscous stress dominated case).

5.5.4.1 Elastic Stress Dominated Case: $\nu < 0.5$

Flows with parameter values, $\nu < 0.5$ (or the elastic stress dominated case), are those associated with higher concentration of polymers per unit volume. The contours of the first invariant, δ_1, for the Zimm's model and the Rouse model, are shown in Figs. 5.7 and 5.8, respectively, at equal snapshots. At lower values of elastic relaxation (represented by the parameter, $We = 15$), the flow-macrostructure evolution in the Rouse model is slow, in comparison with the Zimm's model (i. e., comparing the contour values in the first two rows in Fig. 5.7 versus the first two rows in Fig. 5.8). A reversal of this trend occurs at higher values of elasticity (i. e., the third and the fourth rows in Figs. 5.7 and 5.8 at $We = 20$). We attribute this observation due to the following reasoning. The Rouse model represents 'thicker' fluid, or fluids with slower diffusion than the Zimm's solution, due to the smaller order of the fractional time-derivative. Flows with smaller time-derivative (or the thicker polymer melt case) are those associated with higher concentration of polymers per unit volume [48]. Physically, the formation of these 'spatiotemporal macrostructures' are associated with the entanglement of the polymer chains at microscale [4], leading to localized, non-homogeneous regions with higher viscosity. Experiments [8] have shown that non-Newtonian fluids with a higher polymer concentration, have a greater tendency for the polymer strands to agglomerate (especially at higher elastic relaxation), or the so-called 'over-crowding effect' [49].

Further, notice the disappearance of the structures at larger Reynolds number (i. e., notice the time-snippets at $Re = 1000$, the second and the fourth rows in Figs. 5.7 and 5.8). This observation can be attributed to the fact that the macrostructures are 'washed out' of the channel at higher flow velocities. In an earlier work, we indicated the existence of 'temporally stable regions at high fluid inertia' through a spatiotemporal linear stability analysis [33]. In this work, we associate temporally stable regions with regions devoid of flow structures

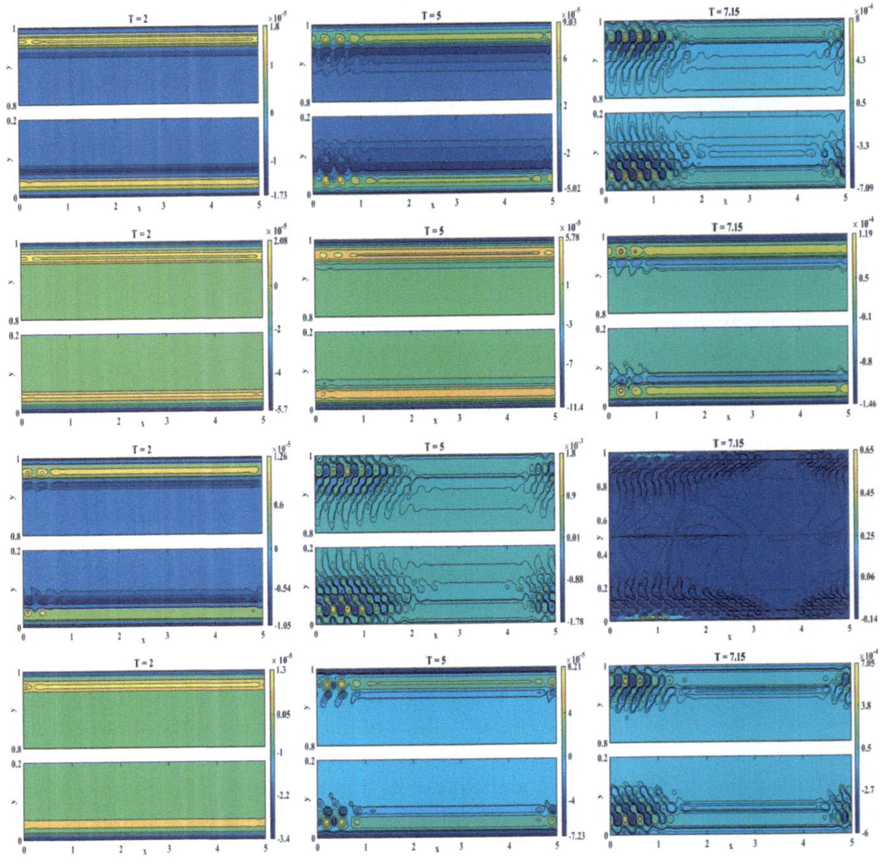

Fig. 5.7 Contours of instantaneous volume ratio, δ_1, for the elastic stress dominated Zimm's model ($\nu = 0.3$, $\alpha = 2/3$). Other parameters set at $Re = 70$, $We = 15$ (first row), $Re = 1000$, $We = 15$ (second row), $Re = 70$, $We = 20$ (third row) and $Re = 1000$, $We = 20$ (fourth row). Plots shown with broken y-axis do not show any structure formation around the centerline

5.5.4.2 Viscous Stress Dominated Case: $\nu \geq 0.5$

Observe that the 'flow structures' for the Zimm's model in the elastic stress-dominated case ($\nu = 0.5$, Fig. 5.9) are larger in size as well as in magnitude, in comparison with the Rouse model (Fig. 5.10), at equal simulation times. Even within the respective models, we find that the macrostructures are more prominent (both in size and magnitude) at higher values of elastic relaxation (i. e., at $We = 20$, third and fourth rows in Figs. 5.9 and 5.10). Again, note that the structure formation is conspicuously absent at larger values of Re (the second and the fourth rows in Figs. 5.9 and 5.10).

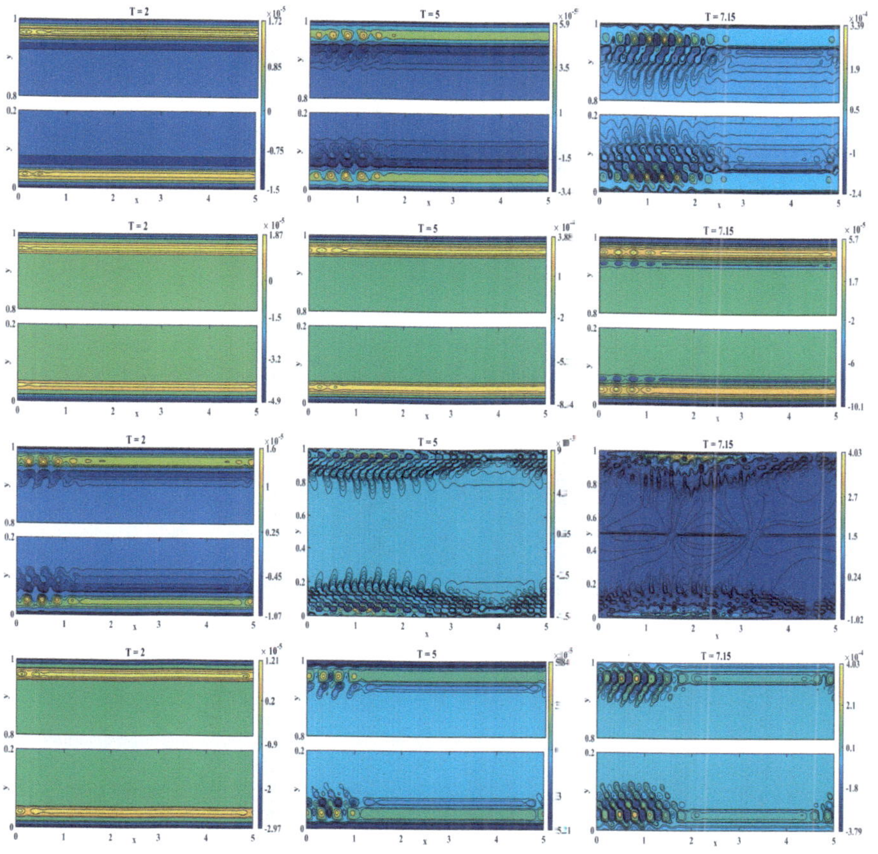

Fig. 5.8 Contours of instantaneous volume ratio, δ_1, for the elastic stress dominated Rouse model ($\nu = 0.3$, $\alpha = 1/2$). Other parameters set at $Re = 70$, $We = 15$ (first row), $Re = 1000$, $We = 15$ (second row), $Re = 70$, $We = 20$ (third row) and $Re = 1000$, $We = 20$ (fourth row). Plots shown with broken y-axis do not show any structure formation around the centerline

We summarize our discussion by noting that for the selected values of the parameters, ν, α, Re and We: (a) The elastic stress-dominated case ($\nu = 0.3$ case) is comparatively more unstable than the viscous stress dominated case, (b) The Rouse model is comparatively more unstable than the Zimm's model at low fluid inertia (or low values of Re) and higher elastic relaxation (or high values of We), and (c) Temporal stability is achieved at higher values of Re, irrespective of the model or the polymer concentration. Thus, our in silico studies not only corroborate the existence of the previously established temporally stable region at high fluid inertia [33], but also highlight the potential of fractional partial differential equations in effectively capturing the flow-instability transition in sub-diffusive flows.

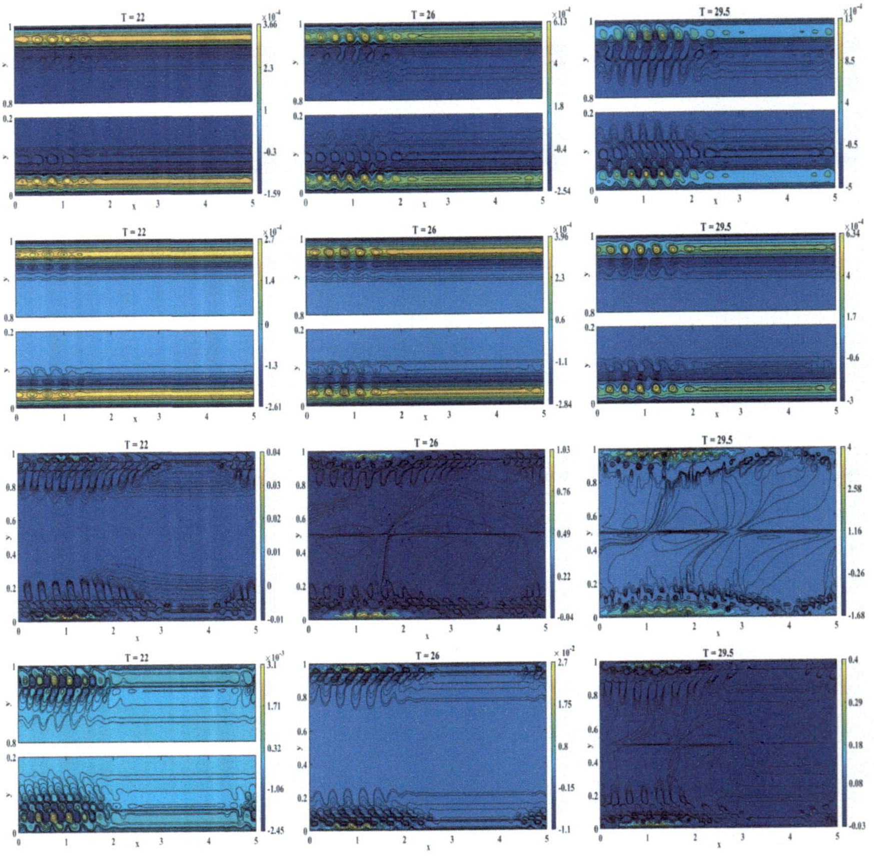

Fig. 5.9 Contours of instantaneous volume ratio, δ_1, for the viscous stress dominated Zimm's model ($\nu = 0.5$, $\alpha = 2/3$). Other parameters set at $Re = 70$, $We = 15$ (first row), $Re = 1000$, $We = 15$ (second row), $Re = 70$, $We = 20$ (third row) and $Re = 1000$, $We = 20$ (fourth row). Plots shown with broken y-axis do not show any structure formation around the centerline

5.6 Conclusions

This investigation addresses the development and numerical simulation of the macroscopic model (5.31), which is upscaled from the microscale model for non-Markovian continuous-time random walk for anomalous diffusional flow-fields relevant for 'thick flows' including polymer melts [33], flows of liquid crystals [34] as well as biological flows including mucus [35] and cartilage [36]. Section 5.2 outlined the details of the numerical method. The numerical results for the sub-diffusive dynamics of the viscoelastic channel flows at low Weissenberg numbers, for the well-known Zimm and Rouse model, are delineated in Sect. 5.5.

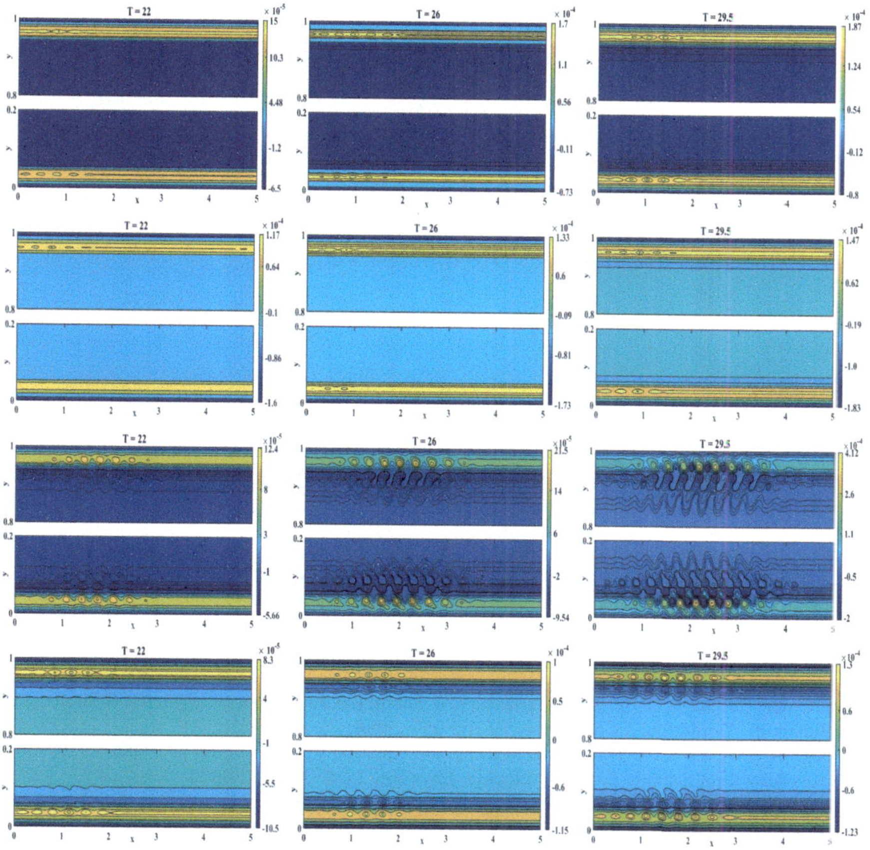

Fig. 5.10 Contours of instantaneous volume ratio, δ_1, for the viscous stress dominated Rouse model ($\nu = 0.5$, $\alpha = 1/2$). Other parameters set at $Re = 70$, $We = 15$ (first row), $Re = 1000$, $We = 15$ (second row), $Re = 70$, $We = 20$ (third row) and $Re = 1000$, $We = 20$ (fourth row). Plots shown with broken y-axis do not show any structure formation around the centerline

5.7 Problems

1. Use the θ-method (developed in Sect. 5.2.1) for spatial (or x-) derivatives, and perform the Von Neumann stability analysis of the following fractional partial differential equation, utilizing the Caputo fractional derivative,

$$z_{ttt}(t,x)+z_{tt}(t,x)+{}_0^C D_t^{\alpha} z(t,x)+z(t,x)=z_{xx}(t,x)+f(t,x),$$
$$0<x<L,\ 0<t<T,\ 0\leq\alpha<1,$$
$$z(0,x)=\sigma_1(x),\quad z_t(0,x)=\sigma_2(x),\quad z_{tt}(0,x)=\sigma_3(x),\quad 0\leq t\leq T,$$
$$z(t,0)=z(t,L)=0,\quad 0\leq x\leq L,$$

where $\sigma_i(x)$, $i = 1, 2, 3$ and $f(t,x)$ are known functions, and $L, T > 0$.

2. Find the dispersion relation, the phase velocity and the group velocity for the time fractional, fifth order, linearized Burger's equation.

References

1. Goychuk I, Póschel T (2021) Fingerprints of viscoelastic subdiffusion in random environments: revisiting some experimental data and their interpretations. Phys Rev E 104(3):034125
2. Lai SK, Wang YY, Cone R, Wirtz D, Hanes J (2009) Altering mucus rheology to solidify human mucus at the nanoscale. PLoS ONE 4(1):e4294
3. Coffey WT, Kalmykov PY, Waldron J (2004) The Langevin equation: with applications to stochastic problems in physics. In: Chemistry and electrical engineering. World Scientific, p 14
4. Rubenstein M, Colby RH (2003) Polymer physics. Oxford University Press
5. Levine AJ, Lubensky TC (2001) Response function of a sphere in a viscoelastic two-fluid medium. Phys Rev E 63(4):041510
6. Kremer K, Grest GS (1990) Dynamics of entangled linear polymer melts: a molecular-dynamics simulation. J Chem Phys 92(8):5057–5086
7. Kou SC, Xie XS (2004) Generalized Langevin equation with fractional gaussian noise: subdiffusion within a single protein molecule. Phys Rev Lett 93(18):180603
8. Fogelson AL, Neeves NB (2015) Fluid mechanics of blood clot formation. Ann Rev Fluid Mech 47:377–403
9. Fricks J, Yao L, Elston TC, Forest MG (2009) Time-domain methods for diffusive transport in soft matter. SIAM J Appl Math 69(5):1277–1308
10. Morgado R, Oliveira FA, Batrouni GG, Hansen A (2002) Relation between anomalous and normal diffusion in systems with memory. Phys Rev Lett 89(10):100601
11. Vainstein MH, Lapas LC, Oliveira FA (2008) Anomalous diffusion. arXiv:0805.0270
12. Gemant A (1936) A method of analyzing experimental results obtained from elasto-viscous bodies. Physics 7(8):311–317
13. Scott-Blair GW (1944) Analytical and integrative aspects of the stress-strain-time problem. J Sci Instr 21(5):80–84
14. Caputo M (1967) Linear models of dissipation whose Q is almost frequency independent-ii. Geophys J Intern 13(5):529–539
15. Bagley RL, Torvik PJ (1983) A theoretical basis for the application of fractional calculus to viscoelasticity. J Rheol 27(3):201–210
16. Rouse PE (1953) A theory of the linear viscoelastic properties of dilute solutions of coiling polymers. J Chem Phys 21(7):1270–1272
17. Tan W, Xu M (2002) Plane surface suddenly set in motion in a viscoelastic fluid with fractional Maxwell model. Acta Mech 18(4):342–349
18. Qi M, Xu M (2009) Some unsteady unidirectional flows of a generalized Oldroyd-B fluid with fractional derivative. Appl Math Model 33:4184–4191
19. Fetecau C, Kamran M, Vieru D (2009) Exact solutions for the flow of a generalized Oldroyd-B fluid induced by a constantly accelerating plate between two side walls perpendicular to the plate. J non-Newt Fluid Mech 189–201

20. Zheng L, Liu Y, Zhang X (2012) Slip effects on MHD flow of a generalized Oldroyd-B fluid with fractional derivative. Nonlinear Anal RWA 13:513–523
21. Zhao J, Zheng L, Zhang X, Liu F (2016) Unsteady natural convection boundary layer heat transfer of fractional Maxwell viscoelastic fluid over a vertical plate. Int J Heat Mass Trans 47:760–766
22. Ancey C (2019) Bedload transport: a walk between randomness and determinism. Part 1. The state of the art. J Hydrau Res 58
23. Siedlik MJ, Nelson CM (2015) Regulation of tissue morphodynamics: an important role for actomyosin contractility. Curr Opin Genet Dev 32:80–85
24. Zaks MA, Nepomnyashchy A (2018) Sub-diffusive and super-diffusive transport in plane steady viscous flows. PNAS 116(37):18245–18250
25. Khalid M, Chaudhary I, Garg P, Shankar V, Subramanian G (2021) The centre-mode instability of viscoelastic plane Poiseuille flow. J Fluid Mech 915(A43)
26. Brunner H, Ling L, Yamamoto M (2010) Numerical simulations of 2D fractional subdiffusion problems. J Comput Phys 229(18):6613–6622
27. Murio DA (2008) Implicit finite difference approximation for time fractional diffusion equations. Comput Math Appl 56(4):1136–1145
28. Jannelli A (2022) Adaptive numerical solutions of time-fractional advection-diffusion-reaction equations. Commun Nonlinear Sci Num Sim 105:106073
29. Crouzeix M (1980) Une méthode multipas implicite-explicite pour l'approximation des équations d'évolution paraboliques. Numer Math 35:257–276
30. Ascher UM, Ruuth SJ, Wetton BTR (1995) Implicit-explicit methods for time-dependent partial differential equations. SIAM J Num Anal 32:797–823
31. Al-Khaled K, Momani S (2005) An approximate solution for a fractional diffusion-wave equation using the decomposition method. App Math Comput 165(2):473–483
32. Podlubny I (1999) Fractional differential equations. Academic Press, p 198
33. Chauhan T, Bansal D, Sircar S (2023) Spatiotemporal linear stability of viscoelastic subdiffusive channel flows: a fractional calculus framework. J Eng Math 141(8)
34. Sircar S, Li J, Wang Q (2010) Biaxial phases of bent-core liquid crystal polymers in shear flows. Commun Math Sci 8(3):697–720
35. Sircar S, Roberts AJ (2016) Surface deformation and shear flow in ligand mediated cell adhesion. J Math Biol 73(4):1035–1052
36. Sircar S, Aisenbrey E, Bryant SJ, Bortz DM (2015) Determining equilibrium osmolarity in poly (ethylene glycol)/chondrotin sulfate gels mimicking articular cartilage. J Theo Biol 364:397–406
37. Singh S, Bansal D, Kaur G, Sircar S (2020) Implicit-explicit-compact methods for advection diffusion reaction equations. Comput Fluids 212:104709
38. Diethelm K, Freed AD (2006) An efficient algorithm for the evaluation of convolution integrals. Comput Math Appl 51(1):51–72
39. Sengupta TK, Sircar SK, Dipankar A (2006) High accuracy schemes for DNS and acoustics. J Sci Comput 26(2):151–193
40. Jin B, Li B, Zhou Z (2018) Numerical analysis of nonlinear subdiffusion equations. SIAM J Appl Math 56(1):1–23
41. Jin B, Li B, Zhou Z (2019) Subdiffusion with a time-dependent coefficient: analysis and numerical solution. Math Comput 88(319):2157–2186
42. Visbal MR, Gaitonde DV (2002) On the use of higher-order finite-difference schemes on curvilinear and deforming meshes. J Comput Phys 181(1):155–185
43. Sircar S, Bansal D (2019) Spatiotemporal linear stability of viscoelastic free shear flows: dilute regime. Phys Fluids 31(8):084104
44. Zimm BH (1956) Dynamics of polymer molecules in dilute solution: viscoelasticity, flow birefringence and dielectric loss. J Chem Phys 24(2):269–278
45. Sureshkumar R, Beris AN, Handler R (1997) Direct numerical simulations of the turbulent channel flow of a polymer solution. Phys Fluids 9:743–755

46. Vaidyanathan T, Robert A, Brasseur JG, Collins L (2006) An improved algorithm for simulating three-dimensional viscoelastic turbulence. J non-Newt Fluid Mech 140:2–22
47. Dubief Y, Terrapon VE, White CM, Shaqfeh ESG, Moin P, Lele SK (2005) New answers on the interaction between polymers and vortices in turbulent flows. Flow Turbul Combust Former Appl Sci Res 74:311–329
48. Bird R, Armstrong R, Hassager O (1987) Dynamics of polymeric liquids. Wiley
49. Doi M, Edwards SF (1986) The theory of polymer dynamics. Clarendon Press, New York, p 391

Chapter 6
Limitations and Future Directions

Abstract Fractional derivatives (FDs) play a crucial role in representing non-local phenomena in both spatial and temporal domains within physics. In space, the long-range effects of the flow are captured by the algebraic decay of the FD kernel. Additionally, FDs, in time, account for non-Markovian effects and thus provide a means to incorporate memory. These effects, spatial non-locality and memory have been observed in diverse fluid mechanical and physical contexts. In this concluding chapter, we explore the prospective future applications and challenges stemming from using FD-based models. It sheds light on how FD-based models could offer valuable insights into various physical phenomena. Moreover, this chapter emphasizes the current limitations and pitfalls in the existing theoretical foundation based on FDs. This comprehensive assessment aims to provide a roadmap for further research and applications of FD models in advancing our general understanding of complex physical systems.

6.1 Introduction

The conventional viscoelastic rheological models are constructed by combining ideally elastic springs with viscous dashpots, e.g., the Maxwell, Kelvin-Voigt, and Oldroyd-B models. These models are rooted in the classical understanding of derivatives with integer orders (i.e., ordinary derivatives). However, when it comes to capturing the intricate dynamic behavior of polymers, particularly slow diffusion, it has been discovered that these traditional models often fall short in accuracy and predictive capability. It is expected to have some discrepancy between experimental results and predictions from viscoelastic models. Researchers often use empirical fitting and adjustments to improve the match between experimental data and model predictions. One possible improvement is to employ more advanced models, such as nonlinear viscoelastic models, to capture complex material behavior better. In many experiments, it is seen that the relationship between the resulting stress and the applied strain obtained from relaxation experiments matches well with FDs of applied stress of order α, where $0 < \alpha < 1$. Furthermore, the diffusion of the fluid

H. Wilson et al., *Viscoelastic Subdiffusive Flows*, Fluid Mechanics and Its Applications 138, https://doi.org/10.1007/978-981-96-0135-6_6

particle follows Fourier's law, in which the mean square displacement follows linear relation $\langle r^2 \rangle \propto t$, with r and t being the displacement and time, respectively. However, the straightforward linear scaling applicable to Fourier diffusion does not hold in many physical systems, i.e. the mean squared displacement increases as $\langle r^2 \rangle \propto t^\alpha$ where the exponent $\alpha \neq 1$. These observations have triggered the development of fractional derivative (FD) models. In the past two decades, FD models have earned a well-deserved reputation as highly effective tools for accurately modeling viscoelastic materials [1–3]. This capability has been demonstrated in previous chapters, where we observed how FD models elucidate the slow-diffusive mechanics of thick viscoelastic fluids, shedding light on several complex phenomena within such flows [4, 5].

The FD models captured many notable phenomena involving non-local, memory-dependent, and anomalous diffusion behaviors appearing in complex systems [6]. Although FD models have many limitations, there are several problems in science and technology in which FD models may help to gain more understanding of flow behavior and develop better technology for the industries. Moreover, the FD-based models are very useful in medical fields, such as understanding mucus flow, blood flow, enhanced oil recovery, and ecology, to name a few. In this chapter, we discuss many limitations of FD models (Sect. 6.2). Motivated by the viscoelastic sub-diffusive flows, a plethora of future problems are presented in Sect. 6.3. We are not limited to only viscoelastic flows. Instead, we give various scenarios and set-ups wherein FD models can provide better explanations and fit experiments. Moreover, the techniques established in preceding chapters for the viscoelastic flows can be adapted and tailored to address specific issues and physical phenomena within diverse problem domains.

6.2 Limitations

Fractional models involving fractional partial differential equations (FPDEs) introduce a rich set of mathematical tools to model complex phenomena that exhibit non-local, hereditary, or anomalous behavior. However, they come with their own set of limitations and challenges [7], which are discussed below.

- The use of fractional models in physics and engineering can raise concerns about their physical consistency. FDs do not have straightforward physical interpretations in the same way that integer-order derivatives do. As a result, researchers may sometimes propose interpretations that do not accurately represent the underlying physical behaviour, leading to questions about the validity of such models in certain contexts [8, 9]. For instance, in fractional models, the units of the parameters often lack a direct physical meaning. Unlike traditional models, where parameters might represent physical quantities like velocity or mass, fractional model parameters may not have clear, intuitive interpretations in terms of real-world phenomena.

This can make it challenging to relate these parameters to specific physical properties or processes, making the models less intuitively understandable in a physical context [10–12].

- A key challenge in using fractional models lies in the proper and well-defined initialization of these models. Unlike traditional integer-order models that rely on initial conditions at a specific time point, fractional models require consideration of all past data from negative infinity to the present moment. This essentially means that fractional models demand an infinite memory of past data, making their initialization computationally demanding and practically impossible in many real-world scenarios. Consequently, dealing with the infinite-memory requirement can be a significant hurdle when applying fractional models to practical problems [13].
- The formulas for FDs are often complex and involve special functions like the Gamma function, Mittag-Leffler function, or others [14–16]. These can be computationally expensive to evaluate. For instance, (i) the Gamma function is defined as an integral that involves an infinite series, making it computationally demanding to evaluate, and (ii) the Mittag–Leffler function, is defined recursively, which can lead to an accumulation of error over multiple iterations, making accurate computation more challenging [17]. Some special functions may exhibit oscillatory behavior or rapid changes in magnitude, which can pose challenges for numerical stability.
- In contrast to conventional differential operators, which often carry specific geometric interpretations related to the behavior or shape of functions (such as their trends or convexity), the fractional derivative does not possess such inherent geometric meaning [18, 19]. This lack of clear geometric meaning can pose challenges in terms of intuitive understanding and interpretation. For instance, developing intuitive strategies for problem-solving or devising new modeling approaches may require more effort and creativity. Furthermore, in practical applications, interpreting the outcomes of models involving FDs may not be as straightforward. Understanding how changes in the fractional order impact the system behavior may require a more abstract and analytical approach.
- Notably, there are more than 30 different definitions of the fractional differential operator [20, 21], reflecting the complexity and variability in approaches to fractional calculus [22, 23]. While fractional integration is relatively well-established and has clear mathematical definitions, the same can not be said for fractional differentiation. The lack of a universally accepted definition for FDs contributes to the challenge of using fractional calculus in various applications. This can create confusion and hinder the development of reliable numerical methods. In addition, different definitions can lead to variations in the behaviour and interpretation of fractional operators, making it important to specify carefully which definition is being used when working with fractional models.
- The use of the Riemann-Liouville (RL) derivative in modeling real-world phenomena with fractional differential equations comes with certain drawbacks. Notably, the RL fractional derivative assigns a nonzero derivative to a constant, posing challenges in specific contexts. Furthermore, if an arbitrary function remains constant at

the origin, the RL derivative displays a singularity at the origin, restricting its suitability for functions such as exponential and Mittag–Leffler functions. Conversely, Caputo's derivative necessitates higher conditions of regularity for differentiability, as it demands the computation of the first-order derivative prior to calculating the fractional derivative. As a result, Caputo derivatives are specifically defined for differentiable functions, whereas RL derivatives can be employed for functions lacking a first-order derivative but possessing FDs of orders less than one [24]. The Caputo–Fabrizio derivative exhibits a localized kernel, and its derivative, when α is set to 0, does not reproduce the initial function. Furthermore, the corresponding anti-derivative lacks a fractional nature [25].

- The definition of fractional integration typically involves a singular kernel [22, 23]. This singularity can lead to complications when solving or simulating fractional-order differential equations. Singular kernels can make numerical computations challenging, as they require special techniques to handle these singularities. What's more, the existence of singularities can influence the stability and convergence of numerical methods, rendering the solution of FD equations more challenging compared to their integer-order counterparts [26, 27].
- Unlike classical derivatives, FDs often lack closed-form analytical solutions, leading to numerical methods as the one and only choice to find approximate solutions. Numerical methods for approximating FDs may encounter numerical instability, especially when dealing with functions that have rapid changes or singularities. Defining appropriate initial conditions as well as boundary conditions for FDs can be quite challenging. These conditions may need to take into account the behavior of the function over the entire domain, hence require computational time and memory. Choosing the right numerical scheme for a given problem can be challenging. Moreover, FPDEs are stiff equations, which pose additional care while handling numerically.
- The models based on FPDEs are relatively unfamiliar in comparison to that of the classical PDEs. Consequently, there is generally a lack of past data and well-known theory for solving them. This can make it hard to apply FPDEs-based models to practical engineering and scientific problems.
- Conducting experiments to verify models based on FDs can be more challenging than for ordinary derivatives. It may require specialized experimental setups or data analysis techniques.
- Although a prominent application of sub-diffusive flows involve modeling polydisperse entangled polymer melts, the framework described in this book is applicable for modeling only monodisperse melts. However, this modeling limitation can be quickly offset by introducing a sub-diffusive tube-based constitutive equation for polydisperse entangled polymers, recently developed by Keunings et al. [28] for classical viscoelastic fluids, which can account for the major non-linear phenomena including reptation, contour-length fluctuations, thermal constraint release, convective constraint release, chain stretch effects as well as the coupling of these effects.

The subject is continually changing, and researchers are actively attempting to create effective and reliable numerical methods for computing FDs. Notwithstanding these challenges, FDs have proven valuable across diverse domains such as physics, engineering, and biology. This highlights the importance of addressing and overcoming these computational obstacles.

6.3 Future Applications

Fractional partial differential equations find applications across diverse fields due to their ability to model phenomena with non-local and memory-dependent behavior [29, 30]. They describe wave propagation in fractal media, viscoelasticity, and anomalous diffusion in physics [31]. Modeling of sub-diffusion and super-diffusion processes is one of the subfields. In biology, FPDEs simulate long-range interactions in processes including population dynamics, tumor growth, and cell migration. They are also employed in finance to simulate complicated systems that exhibit memory effects in risk management, portfolio optimization, and option pricing [32]. FPDEs can also be used to model heat conduction in non-homogeneous materials, describe fluid flow in porous media, and analyze structural mechanics with memory effects, among other engineering applications [29, 33]. Their versatility makes FPDEs a powerful tool for tackling complex, real-world problems in various scientific and engineering disciplines, encompassing subfields like fractional diffusion, fractional wave equations, and more specialized domains within each of these areas [34, 35]. This section provides future research problems–covering broad applications from physics, chemistry, biology, and engineering—which can be modelled through FPDE models. Note that not all research topics listed here involve viscoelastic fluids. Nevertheless, FPDE techniques and ideas developed in previous chapters can be easily extended to other complex fluids.

6.3.1 Enhanced Oil Recovery/Polymer Flooding

Given the escalating global demand for petroleum, optimizing the efficiency and cost-effectiveness of oil fields has become an increasingly pressing concern. This urgency is accompanied by the predominance of mature oil fields and the declining rate of discoveries. Consequently, enhanced oil recovery (EOR) techniques have gained paramount significance [36, 37]. Polymer flooding technology is the preeminent method for chemical EOR [38]. Its proficiency lies in its capacity to attenuate the water-oil mobility ratio, thereby augmenting the areal and vertical sweep efficiency of the recovery process, see Fig. 6.1. This renders polymer flooding a pivotal tool to extract oil from reservoirs with gained adequacy yet economical. Polymer flooding is a technique employed in specific reservoirs where conventional water flooding may be less effective. These situations include fractures or areas with high permeability,

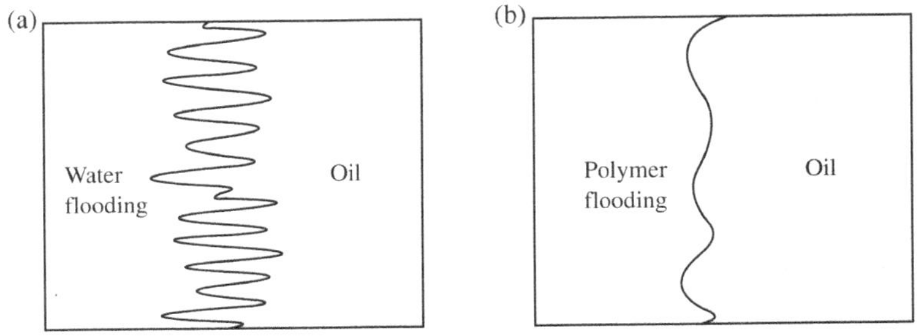

Fig. 6.1 Schematic of Hele-Shaw cell showing **a** water flooding and **b** polymer flooding in the oil reservoir. Enhancing the viscosity of the injected fluid using polymer (polymer flooding) is a strategy to defer water breakthrough, optimize sweep efficiency, and minimize fingering instability

where injected water can be redirected or concentrated. Additionally, it is helpful for heavy oils that are resistant to flow.

In polymer flooding, the viscosity of the displacing solution is increased by mixing a tiny amount of a polymer (approximately 200–1500 ppm). Consequently, it reduces the mobility ratio of water to oil and thus enhances the efficiency of the injected displacing solution [39]. In other words, incorporating a water-soluble polymer into the injected water alters its properties. This modification enables the water to penetrate and traverse a more significant portion of the reservoir rock, leading to a notable increase in the recovered oil proportion [40]. Nevertheless, it should be noted that the polymer flooding does not efficiently produce substantial enhancement in microscopic displacement owing to a relatively minor change in capillary number. Nonetheless, polymer flooding leads to more quickly and cost-effectively residual oil saturation [41].

Polymer flooding has been employed in various trial studies. Nevertheless, it encounters tremendous challenges, notably the pronounced retention of chemicals, which harm the economic viability of the process. Moreover, issues of suboptimal injectivity pose a risk of uncontrolled fracturing, further complicating the successful implementation of this technique [42–44]. Note that polymer retention increases (i) with the molecular weight of the polymer along with the clay content, (ii) with the lowering permeability of the reservoir rock, and (iii) with the weakening anionic charge of the polymer. The three mechanisms mentioned above primarily account for polymer retention in porous media. The key contributor to this retention effect is adsorption, where polymers bind to the surface of the rock [45]. Due to adsorption, polymer concentration dilutes, which in turn reduces the viscosity of the polymer, hence reducing the effect of polymer flooding. Moreover, the slow diffusion rate of polymer molecules hinders the mixing process, extending its impact to the lithofacies scale [44, 46]. It is also important to note that the displacement systems are prone to fingering instability due to differences in mobility. These mobility differences can be due to viscosity or permeability [47]. Many numerical studies showed that control in viscous fingering leads to more oil recovery [48, 49].

To understand fingering instabilities arising in chemically driven EOR, [50] have reported the numerical simulations and linear stability analysis to analyze miscible and immersible viscous fingering in a Hele-Shaw cell. They used the Oldroyd-B model for modeling the viscoelastic fluid. Note that the Oldroyd-B model is valid for dilute polymers. This study confirmed that increasing the elasticity, represented by the Weissenberg number, of the displaced or displacing viscoelastic fluid gives more flow stabilization. A similar analysis has been performed for miscible viscous fingering using the White-Metzner model to model the nonlinear viscoelastic behavior of the polymer fluid. While these dilute viscoelastic models have been extensively studied and found to provide theoretical and numerical frameworks of miscible and immiscible viscous fingering emerging in polymer flooding slow-diffusion of polymer and heavy oil (displacing and displaced fluids), they have been ignored. Nonetheless, the crude oil mimics a thick viscoelastic fluid in tertiary oil recovery, leading to slow diffusion. Note that the diffusion process of viscous or thick oil plays a pivotal role during the tertiary stage of oil recovery with polymer flooding.

A promising direction for future research in Hele-Shaw flow systems involves capturing the viscoelastic sub-diffusive behavior, thereby establishing more intricate model systems. We have seen in previous chapters that the viscoelastic properties of fluid rheology, including sub-diffusivity, can be represented through the FPDEs [4, 5], incorporating FDs [51–53]. It has been shown that the fractional Newton's law of viscosity yields a generalization of classical Darcy's law [51, 54]. In the context of polymer flooding, these enriched FD-based models offer a means to determine effective flow parameters, to understand viscous fingering instability, to understand the role of various physical effects, including the retention effect, and to explain the impact of slow diffusion and adsorption.

6.3.2 *Viscoelastic Nanofluid*

Nanofluid refers to a fluid that comprises particles of nanometer scale, commonly referred to as nanoparticles. Nanofluids have attracted the attention of scientists and engineers for decades [55]. Thanks to their potential to enhance the thermal conductivity of fluids such as water, synthetic oil, ethylene glycol, pump oil, and motor oil. Thus, the nanofluids help to improve the heat transfer characteristics. Among all types of nanofluids, viscoelastic nanofluids have several industrial applications, e.g. heat exchangers, cooling of automobile motors, welding machines, energy devices, and micro-electromechanical systems. It should be noted that slow diffusion often occurs in viscoelastic nanofluids [56]. Adding nanoparticles to a viscoelastic base fluid can change its diffusion properties, resulting in a slower diffusion rate than a base fluid without nanoparticles [57, 58]. This behavior can be influenced by factors such as the type and concentration of nanoparticles and the specific rheological properties of the viscoelastic nanofluid.

Using FDs in modeling viscoelastic nanofluids contributes to several compelling improvements. As discussed earlier, FDs effectively capture non-local and memory-dependent behavior, thus enhancing the response of the material to applied forces or changing conditions. Through FDs, the transient behavior of viscoelastic nanofluids can be predicted more accurately. Additionally, FDs address anomalous diffusion, which may occur in some nanofluids. These derivatives also offer flexibility in modeling complex rheological properties, allowing for the incorporation of memory effects and better representation of deformation or flow responses. FDs facilitate the optimization of nanoparticle concentration and control in manufacturing processes. This approach contributes to advancements in scientific understanding and opens up innovative applications in several engineering applications. Employing FDs in viscoelastic nanofluid modeling significantly enhances accuracy. It improves performance and applications in various areas, including enhanced oil recovery, drug delivery systems, environmental remediation processes, tissue engineering, the food and beverage industry, and biotechnology.

The research on viscoelastic nanofluid is still in its early stages. A few numerical studies employed nonlinear fractional boundary layer equations to determine heat transfer in viscoelastic nanofluids. For instance, the unsteady boundary layer flow and heat transfer of a nanofluid with fractional Maxwell viscoelastic properties over a moving plate were investigated by [59]. They modeled nanofluidic equations using the nonlinear fractional boundary layer equations and the time-dependent FDs for the convection terms. The velocity and temperature relaxation time enhanced the convection flow and heat transfer well. In contrast, the velocity and temperature FD parameters have the opposite effect. In another study, [60] explored the incompressible, steady, and laminar fluid boundary flow and heat transfer across a moving plate. Their investigation involved a distinctive set of constitutive relationships, integrating relaxation time parameters and distributed order time fractional operators. These relationships governed the flow and heat aspects, considering relaxation characteristics and a spectrum of memory and nonlocal features. Liu et al. [60] derived numerical solutions by approximating distributed order integrals by summing multi-fractional terms. Additionally, the discretization of FDs was achieved using the L1 and L2 schemes [61]. There is still a relatively limited exploration of modeling viscoelastic nanofluids using FPDEs, indicating a promising area for future investigation.

6.3.3 Granular Flows

Granular materials display complex behaviors originating from particle-particle interactions. These interactions further lead to several phenomena, e.g., the caging effect, clustering, jamming, shear-banding, force-chain, segregation, avalanche formation, oscillons, surface wave, etc. [62–69]. In the caging effect, the particles in a granular material are temporarily trapped by their neighboring particles, leading to erratic movement [65, 66]. The clustering and jamming in granular flows refer to the state where particles tend to cluster together and tend to jam, respectively [67].

Note that the contacts between particles in a granular material are highly nonlinear, leading to complex force networks. Moreover, mixing and segregation is a peculiar behaviour of granular flows [70]. For instance, a rotating drum partially filled with granular materials shows avalanche, continuous surface flow, and segregation [71, 72]. Christov and Stone [73] showed that mixing particles in a rotating drum exhibit self-diffusion of particles, with the axial diffusion being anomalous.

In the case of granular flows, anomalous diffusion occurs when granular particles within a granular material (sand, grains, powder, etc.) move in a manner that deviates from the typical Brownian motion and diffusion behavior observed in gases and liquids [65]. The term anomalous diffusion refers to the behavior of particles or matter in a medium whose movement does not follow a classical, linear, or regular diffusion pattern. This implies that the particle movement cannot be predicted based on standard diffusion equations, and the mean squared displacement of particles adopts a power law in time having an exponent different from unity [73]. The anomalous diffusion has been well reported through experiments and particle simulations for driven granular systems [74, 75].

The experimental investigation of the diffusion properties of a two-dimensional bi-disperse dry granular material subjected to quasi-static cyclic shear is reported by [65]. They analyzed a comparison between the behavior of granular matter with the hard-sphere systems Lennard–Jones fluids, and experiment results of colloidal systems at the glass transition. Despite fundamental microscopic distinctions (such as thermal versus athermal nature), the study reveals significant similarities in their statistical behavior. Their research presented the intricate dynamics of particle confinement—cage dynamics–which govern sub-diffusion in the "slow relaxation" regime. They successfully identified and quantified the admissible physical scales associated with the cage phenomenon.

The diffusion in bi-disperse mixtures with concentration-dependent diffusivity was investigated by [73]. They provided an asymptotic explanation for self-similar behavior, a scenario lacking an exact analytical solution. Their discovery revealed two distinct regimes: one featuring anomalous scaling following experimental observations for poly-disperse granular materials and another demonstrating normal diffusive scaling over extended time intervals. Additionally, [76] analyzed experimental investigations on diffusivity in two-dimensional dense granular Couette flows. Noteworthy was the observation that diffusivities aligned with the mean flow direction were approximately twice as large as those in the perpendicular direction. It was shown that an anisotropic force network introduces an extra level of anisotropy in diffusivity. Note that, this characteristic is unique to dense systems and lacks a clear parallel in rapid granular flows. Specifically, diffusivity experiences a reduction along the axis of the dominant force network. The particle motion appeared sub-diffusive due to the decreasing shear rate away from the shearing surface or super-diffusive due to Taylor dispersion effects.

What follows is the observation that granular diffusion can manifest sub-diffusive or super-diffusive behavior, indicating particles spreading slower or faster than classical diffusion models predict. While sub-diffusion is well-observed experimentally and through particle-level simulations, a comprehensive mathematical theory has not yet been established. The utilization of FPDE-based models in granular rheology contributes to a more accurate mathematical framework for understanding granular diffusion and describing the intricate behavior of granular materials. Furthermore, models based on FPDEs serve as a robust mathematical tool for fitting experimental data derived from granular materials, facilitating the extraction of relevant parameters that enhance the accuracy of describing the diffusion process in various physical and industrial applications. The exploration of fractional granular rheology is entirely untouched, presenting a tremendous opportunity for future research.

6.3.4 Mucus Layer

Mucus layer is a protective wall formed in various body parts, mostly in areas exposed to the external environment or internal organs that come into contact with foreign substances. It comprises glycoproteins, water, salts, and other compounds [77, 78]. Mucus has a complex, heterogeneous polymer scaffold with viscoelastic properties [79]. The mucus layer serves several critical functions in the respiratory and gastrointestinal systems [80]. First and foremost, the mucus layer protects us from foreign particles, such as dust, bacteria, viruses, etc., by trapping and removing these elements before they can reach the underlying cells. In addition, the mucus controls the movement of small-sized ions, nutrients, and hormones and therefore helps to stay healthier [77, 81]. Many components related to the immune system, such as antibodies, antimicrobial peptides, and immune cells, are found in mucus and aid our body in defending against pathogens such as algae, bacteria, fungi, etc. It provides lubrication to prevent tissues from drying out in specific areas like the respiratory tract, where air constantly passes over delicate tissues. Furthermore, in the gastrointestinal tract, the mucus layer helps lubricate food, making it easier to swallow and aiding in food movement through the digestive system [78]. It contains enzymes, which assist in breaking down complex molecules. In the stomach, mucus helps protect the lining from the acidic conditions essential for digestion, thereby maintaining and regulating the pH level, see Fig. 6.2.

Various body regions generate specialized types of mucus tailored to their specific functions. For instance, the mucus in the respiratory passages is typically thinner, aiding in the ease of breathing. On the other hand, in the digestive tract, mucus tends to be denser, assisting in digestion and shielding the mucous membrane from acidic environments (Fig. 6.2). In short, mucus layers are essential for maintaining the health and proper functioning of various organs and systems in the human body. The disruptions in the production or composition of mucus can lead to multiple health issues. For example, excessive mucus production can occur in conditions like chronic bronchitis, while cystic fibrosis can lead to abnormally thick and sticky mucus. Understanding and controlling diffusive transport across mucus layers is pivotal in developing and designing effective drug delivery systems [78, 83].

Diffusive transport of ions, bacteriophages, or other particles within the mucus layer is a well-observed biological phenomenon [79]. Note that bacteriophages, or

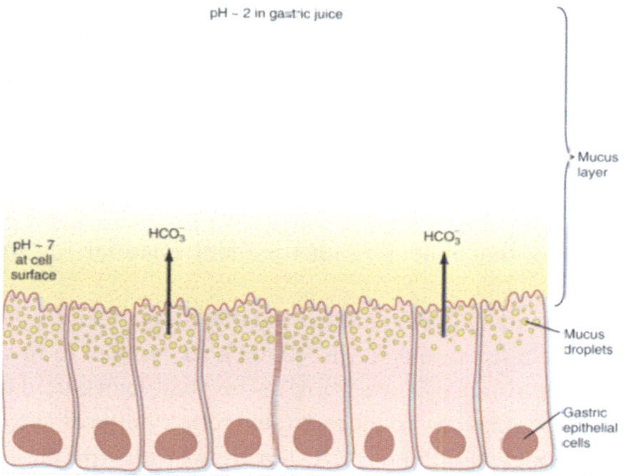

Fig. 6.2 Mucus protects our stomach from its own acid by limiting back-diffusion of acid and proteins, and balancing the pH gradient through their bicarbonate (high near the lumen, low near the epithelial cells). *Source* [82]

phages, are specialized viruses that target and eliminate bacteria. Without independent mobility, phages rely on diffusion to locate and infect bacterial hosts. Barr et al. [84] revealed a notable phenomenon: A phage with weak adhesive properties to mucus manifests sub-diffusive motion, in contrast to the typical pattern of normal diffusion, when traversing mucosal surfaces. Through theoretical models and factual experiments, they illustrate that the sub-diffusive motion significantly enhances phage encounters with bacteria, particularly in scenarios of low bacterial concentration. It is worth noticing that this represents a unique instance where a predator employs a sub-diffusive search strategy effectively.

The barrier characteristics of gastric mucus in the context of the digestive system have been a subject of prolonged interest for decades [85]. As the mucus layer lies next to the epithelial surface of the stomach, separates the stomach lining (pH ≈ 7) from the acidic environment of the gastric lumen (pH ≈ 2), see Fig. 6.2; this significant difference in pH levels is crucial for proper gastric operation. However, this gradient is maintained over a relatively small distance (typically around 500 μm thick for the gastric mucus layer). This has led some to characterize the mucus layer as a barrier to diffusion [85]. Several experiments have been reported on [e.g. 86–89], to understand various behaviors of mucus and their significance on the diffusion rate of small ions through hydrated mucus gel. There are many conflicts in experimental results [86–89]. To understand the conflicting experimental data, a number of modelling strategies have been proposed. Recently, [90] analyzed a Nernst–Planck-like model of electro-diffusion of cationic elements (e.g., hydrogen protons) through a charged chemically reactive domain between two bathing environments. They aimed to mimic a lab experiment called a "flux chamber" experiment, similar to what studies by [87, 88]. They examined the attachment and detachment of the cation to the surrounding gel and investigated the impact of the Donnan electric potential at the a gel and a bath interface. Additionally, the study included findings on the transport and diffusion of small ionic substances within mucus gels.

A similar model was used by [91], which showed that the mucus layers slow the hydrogen diffusion. However, when hydrogen chemically binds with the network, this effect gets weaker. Ernst et al. [92] studied a model known as the Brownian diffusion model that looks at the structure of cells with permeable membranes to explain why nanoparticles in mucus move slower than expected. This model is based on Brownian diffusion of particles in a confined geometry, made from permeable membranes. Note that numerical simulation accurately predicted the observed behavior in experiments, showing that mucus is an excellent material for studying transient sub-diffusive regime.

The mucus layer can be modelled using FPDEs. Following the work of [91], one can use a fractional Nernst–Planck-like model to investigate sub-diffusive phenomena, for instance. The FD proves invaluable in elucidating diffusion within the mucus layer. Its capacity to incorporate memory effects, account for non-local interactions, and model anomalous diffusion aligns well with the complex dynamics of biological systems. This flexibility allows us to capture the specific characteristics of gastric mucosa and other mucus layers, offering a more accurate and clear representation of sub-diffusion processes in the critical biological context. Considering insights from FPDEs may provide a good fit for experimental data. Given that the mucus layers are often heterogeneous, with variations in structure and composition, FDs provide a means to consider and incorporate heterogeneity, thus allowing for a more realistic scenario. To what follows, FPDEs enable us to capture the unique characteristics of mucus and offer a more accurate representation of particle behavior in the complex biological environment of the mucus layer.

6.3.5 Liquid Crystals

Liquid crystal (LC) exhibits properties of both liquids and crystalline solids. Sometimes, LCs are called the fourth state of matter or mesophase. The molecules are ordered like a crystal in a liquid crystal but can still flow like a liquid. In other words, they carry a liquid-like flowing nature, but because of their positional order, such compounds are often more viscous. This liquid-like nature gives them unique optical, electrical, and mechanical properties. There are several types of LCs, including nematic, smectic, and cholesteric phases, each with distinct arrangements of molecules [93, 94]. Nematic LCs (e.g., rod-like or disc-like) are characterized by lower order than smectic phases. In nematics, the molecules' centers of mass do not have a fixed arrangement (no positional order). It is similar to how molecules behave in isotropic fluids. However, the relatively long molecules align directionally and strive to stay parallel in localized regions [93, 95]. In the nematic phase, the molecules are pointed in the same direction (orientational order) and thus free to rotate or slide past one another. The molecules keep aligned with the nematic phase in the smectic phase but are additionally organized in layers. There are different types of smectic phases based on the angles formed between the molecular axes and the planes of molecules. In the cholesteric phase, the molecules line up in a specific direction and are arranged in a spiral pattern. Each layer is slightly twisted compared to the layers above and below it. The degree of molecular organization increases from the nematic to the cholesteric phases. LCs find numerous applications owing to their electro-optical properties, for instance, liquid crystal displays (LCDs), optical modulators and shutters, polarizing films, electronic paper displays, phase modulators in optics, variable neutral density filters, light modulation in projectors, switchable lenses in cameras, medical diagnostics applications, energy-efficient windows, LC thermometers, [95].

Mainly, there are two types of continuum theory to understand the behavior of LCs: (i) Ericksen–Leslie theory [96–98] and (ii) Oseen–Frank theory [99, 100]. The former theory considers the director field (which represents the average molecular orientation) and describes the behavior of LCs in terms of their viscoelastic properties, flow, and deformation. The latter theory provides a simplified description of LCs based on their elastic properties and is particularly useful for describing the behavior of nematic LCs. These theories serve as the foundation for modeling and understanding the behavior of LCs in various applications, including LC displays (LCDs), optical devices, and other technologies. They have been refined and extended over the years to accommodate more complex scenarios and phenomena observed in LC systems.

Owing to ordered molecular structure and anisotropic properties, LCs can lead to sub-diffusive behavior [102–104]. The presence of domains [105], defects (singularity) [106, 107], and other structural features within the liquid crystal material can create obstacles that disrupt the movement of particles, resulting in sub-diffusion, see Fig. 6.3. Several molecule-dynamic simulations and experiments have been conducted to observe and quantify anomalous diffusion in LCs and LC collides [101, 102, 108, 109]. Maeda et al. [103] demonstrated that electro-convection in homeotropically aligned nematic liquid crystal systems exhibits distinctive nonlinear phenomena resulting from the interaction between convection and the nematic director, leading to the emergence of Nambu–Goldstone modes. Using the diffusion coefficient, they experimentally analyzed the movement of tagged particles driven by the turbulence. They reported that sub-diffusion occurs in a certain time range, suggesting turbulence causes particles to rebound. Further, the rebounding is induced by characteristic linear structures of the nematic director caused by turbulence. It is worth noticing that sub-diffusion, which results from the interplay between the nematic director and turbulence, is unique to nematic liquid crystals. Furthermore, liquid crystals are prone to show instability and patterns [110], wherein diffusion may not be normal. Additionally, particle transport is influenced by external electric fields, resulting in anomalous diffusion within the LC environment [111]. In addressing this complexity, FPDEs offer a mathematical framework suitable for numerical simulations and experiments. They enable precise predictions and analyses of particle behavior in LCs subjected to electric fields or other external influences [112–114].

One such study was carried out by [115] in which they compared experiments in the background of a coupled system of diffusion and Poisson equations. Their results match well with the nematic and twist-bend phases, indicating sub-diffusive regimes in the system. Note that the FPDE models are helpful because (i) the particle

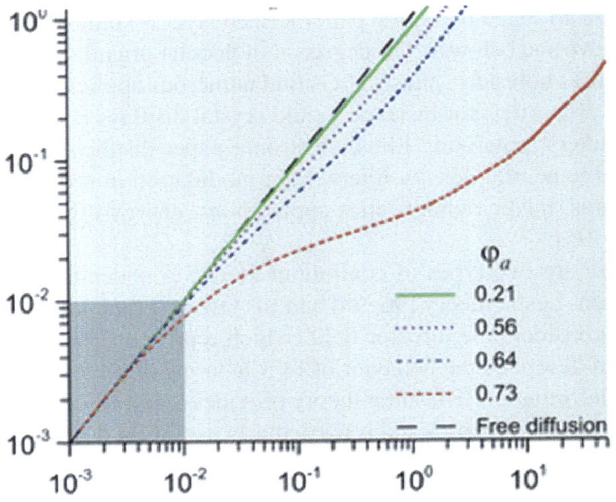

Fig. 6.3 Scaled mean square displacement of colloidal particles in monodisperse suspensions, confined in a quasi-two-dimensional geometry, for various values of the particle area fraction φ_a. The MSD for all systems is practically linear except the system with a large area fraction (red-dashed line), where it shows a sub-diffusive nature. *Source* [101]

movement in an LC is influenced not only by the current electric field but also by past interactions, resulting in memory effects, and (ii) under the influence of electric fields, particle transport can deviate from conventional models leading to anomalous transport and sub-diffusive transport. Moreover, FPDE models can be used to study the long-term effects of electric field-driven particle transport. In particular, this is important for applications where sustained or extended transport is crucial, such as in LC-based devices. Furthermore, a theoretical study of wave motion in viscoelastic phononic crystals based on FD-based models has been conducted in Ref. [2]. They used the fractional Kelvin–Voigt model for viscoelastic material and solved it using a finite-difference scheme. Their finding indicates that strategically selecting host materials with specific viscoelastic properties holds promise for creating phononic crystals with expansive transmission gaps. This innovative approach in material design aligns with the methodologies elucidated in preceding chapters, which can be extended and adapted to explore complex phenomena, including diffusion, within LCs. FPDE models present an exciting approach to advancing our understanding and exploiting the unique properties of LCs for various technological innovations. As researchers continue to unravel the complexities of FDs, we can anticipate breakthroughs in liquid crystal science and its diverse applications.

6.3.6 Ecology

FPDEs can capture a wide range of ecological phenomena due to their ability to model memory effects, non-local interactions, and complex spatial patterns [116–118]. Notably, these models are more appropriate for modeling ecological processes that exhibit anomalous diffusion (e.g., the spread of invasive species, movement of individuals in heterogeneous landscapes, and dispersion of nutrients in soils) and population dynamics with memory effects where past conditions influence current dynamics, e.g., species affected by past disturbances [119–121]. The FDs are more suitable for explaining spatial patterns and patchiness in ecological systems, for instance, the formation of aggregations, clumps, or fractal-like structures in populations and the distribution of resources across a landscape [122]. Ecological processes that involve the dispersal, migration, or movement of organisms can be modelled through these models since FDs account for long-memory effects in movement patterns [123]. Furthermore, FPDEs can also be helpful for understanding (i) foraging behavior in species by which an organism acquires and utilizes sources of energy and nutrients based on the memory of past encounters with resources, (ii) reaction-diffusion systems with non-local interactions [124–126], (iii) the stability and resilience of ecological communities, (iv) transport and transformation of nutrients, pollutants, and other biochemical substances in ecosystems.

In biological models, interaction networks form food chains where species are interconnected through trophic relationships. Phytoplankton, integral to these ecosystems, serve as a vital link by providing sustenance for marine life, generating oxygen essential for humans, and contributing to atmospheric purification through carbon dioxide consumption. However, the accelerated growth of phytoplankton can impact oxygen availability to develop other aquatic plants and animals. Existing literature delves into mathematical models that explore the relationships within these ecosystems, mainly focusing on the interaction dynamics involving nutrients, phytoplankton, and zooplankton. Since biological systems present long-range temporal memory or long-range space interactions, the use of FDs can efficiently handle the dynamics of a disease model and also give information on each model point.

Further, ecological systems are the best examples of sub-diffusion. Traditionally, the prey would exhibit normal diffusion patterns, dispersing evenly across the environment. However, due to specific ecological conditions, the prey's movement deviates from classical diffusion, displaying characteristics of sub-diffusion [127–129]. Several factors contribute to this sub-diffusive behavior. The complex terrain of the habitat, perhaps with uneven topography or obstacles, impedes the free movement of the prey. Additionally, the prey species might remain in specific regions or show slower exploration due to territorial behavior, limited resources, or a heightened awareness of predating risks. As a result, the predators, in turn, adjust their hunting strategies to account for the sub-diffusive movement of the prey. The predators may need to adopt more strategic and patient approaches, relying on a heightened awareness to anticipate the slower spread of prey. This can lead to dynamic shifts in the spatial distribution and interaction dynamics between the predator and prey populations. Mathematically modeling predator-prey systems with sub-diffusion involves incorporating FDs to represent the slower and more localized movement of the prey species accurately. This approach allows ecologists to gain insights into how sub-diffusion influences the spatial dynamics, distribution patterns, and, ultimately, the stability of the predator-prey system within the given ecological context. Incorporating FDs in numerous analogous ecological set-ups to capture various effects, including temporal memory, space interaction, and sub-diffusion, presents a promising future direction.

6.3.7 Plasma

Plasma is a matter characterized by positively and negatively charged particles with roughly equivalent charge densities [130]. It arises when a regular gas is heated to an extreme temperature, causing the molecules to possess kinetic energies surpassing the ionization threshold. Consequently, collisions lead to the detachment of electrons from the atoms, resulting in a blend of ions and free electrons. Since the ionization process typically starts at a well-defined temperature ($\approx$1000 Kelvin), plasma is frequently denoted as the "fourth" state of matter. In most common plasma, the charged particles exist in a state of unbound gas, and thus, the kinetic energy is much greater than the average electrostatic energy [131]. Due to free electrons and ions, plasmas are excellent conductors of electricity. Moreover, plasmas can generate their magnetic fields and attain extraordinarily high temperatures and energies, rendering them crucial in a wide array of high-energy phenomena.

Plasmas are seen in the universe, e.g., stars, sun, earth lightening, and ionized air, produced by high temperatures and electrical discharges. They are seen in various applications, e.g., nuclear fusion, plasma TV screens, fluorescent lights, plasma medicine, astrophysics simulations, etc. Plasmas containing charged solid particles (mono-dispersed micrometer-sized dust grains) and ions and electrons are known as dusty or complex plasmas [132]. In these plasma systems, the surrounding plasma can strongly influence the motion of dust particles, leading to sub-diffusive behavior [133]. The interaction between charged dust particles and the plasma can give rise to complex and nonlinear dynamics. Under the influence of strong magnetic fields, plasmas experience highly constrained motion. Due to magnetic fields, charged particles tend to follow magnetic field lines, which can result in sub-diffusive behavior perpendicular to the magnetic field [134–136]. Qin et al. [134] identified that classical diffusive transport across the mean magnetic field is not maintained in nearly slab magnetostatic fluctuations. Once parallel diffusion sets in, i.e., after particles travel about a parallel scattering mean free path, perpendicular random walk is suppressed to a sub-diffusive level. The sub-diffusive motion is often seen in fusion experiments [137], arising due to complex geometries or irregular boundaries. The occurrence and extent of sub-diffusion in plasma depend on the particular conditions and attributes of the plasma, alongside the influence of external factors such as magnetic and electric effects on complex boundaries.

Interestingly, when describing fusion plasmas, a super-diffusive framework is typically employed. The foundation of complex phenomena like anomalous diffusion often lies in Lévy statistics, which depict fractal processes. In this context, anomalous diffusion was modelled through the fractional Fokker–Planck equation and solved numerically by [138]. In the model, they incorporated fractional velocity derivatives and Langevin dynamics to capture non-local transport Their analysis presented a way to understand the anomalous transport of charged particles in plasmas under a magnetic field.

Studying how molecules move independently in natural fluids is quite tricky in experiments. Scientists often rely on indirect techniques like nuclear magnetic resonance, which cannot track the movement of individual molecules [139]. However, in molecular dynamics simulations, applying the Einstein relation to find the coefficient of self-diffusion is straightforward. Such studies include Lennard–Jones fluids and Yukawa systems. Nunomura et al. [135] performed very low gas pressure experiments on complex plasmas. They predicted the self-diffusion in a monolayer liquid complex plasma and also provided the estimate of the self-diffusion coefficient. In addition, they validated experimental results with simulations. Their experiments ensured that the complex plasma acted like actual solids and liquids, not like tiny particles suspended in a liquid (colloids). This is important because, in real solids and liquids, the movement of particles depends on vibrations called phonons. They showed that near the melting point, the motion was slower than usual (sub-diffusive), but as the temperature increased, it became more regular (diffusive). They also discovered that the self-diffusion coefficient varies linearly with increasing temperature; see Fig. 6.4.

It is important to highlight that in fusion plasmas, spatial non-locality and memory effects are anticipated. Specifically, phenomena like avalanche-like events, streamers, and zonal flows lead to unusually extensive particle displacements, which exhibit characteristics of non-local processes. In contrast, anomalously extensive trapping events by coherent structures, like turbulent eddies, introduce memory effects into the transport process. An example of such an event is plasma turbulence. The turbulence in complex plasmas refers to the chaotic and complex behavior exhibited by the motion of charged particles (ions and electrons) and dust grains within the plasma. In plasmas, the presence of charged dust grains adds an extra layer of complexity to the dynamics. The interaction between the charged particles and the dust grains, along with the collective effects of the particles' motion, can lead to highly irregular and unpredictable flow patterns, and, hence, turbulence. This turbulence can manifest in various ways, such as the formation of structures, fluctuations in density and temperature, and the [140] examined the movement of tracer particles within a model of three-dimensional, resistive, pressure-gradient-driven plasma turbulence. They found that the anomalous transport in this system, see Fig. 6.5. The radial displacements of tracers displayed a probability density function that deviated from Gaussian characteristics, featuring decaying tails and following a super-diffusive

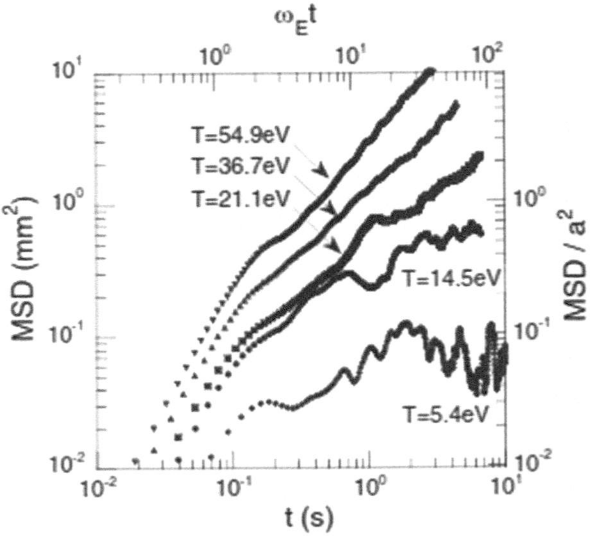

Fig. 6.4 The mean-square displacement (MSD) at various temperatures in the solid and liquid phases. The MSD is proportional to t^2 on a short time scale $\omega_E t \leq 1$, (correspond to ballistic motion of the particles). It increases linearly with time for a long time scale in the liquid phase (top three curves) corresponds to diffusion. For solid phase MSD (lowest curve) is mostly constant at $\omega_E t \geq 1$. A small dip in the MSD slope (corresponding to sub-diffusion) for the temperatures above the melting point ($T \approx 14.5$) at time $t = 0.1$–0.2 s can be seen, indicating caged motion. Here T, ω_E, and a are the temperature, Einstein frequency, and the mean particle separation, respectively. *Source* [135]

scaling pattern for their moments. They utilized a macroscopic transport model for transport of tracer particle and modelled with the FD operators in both space and time. Some studies on turbulence in dusty plasma are reported in [141–144].

Several investigations have demonstrated that employing methods based on FDs provides a comprehensive theoretical framework for gaining profound insights into physical phenomena such as sub-diffusion, anomalous diffusion in dusty plasmas, and turbulence in both regular and fusion plasmas. However, it is crucial to emphasize that, despite their potential, a future challenge exists as the application of FPDE-based models in these contexts has not been thoroughly explored or reported. This highlights a promising area for future research and development.

6.3.8 Blood Flow

Blood is a complex fluid that does not behave like a Newtonian fluid and exhibits non-Newtonian behavior due to its composition comprising cells, plasma, etc. This composition gives blood its unique rheological (flow) properties. Blood can flow and

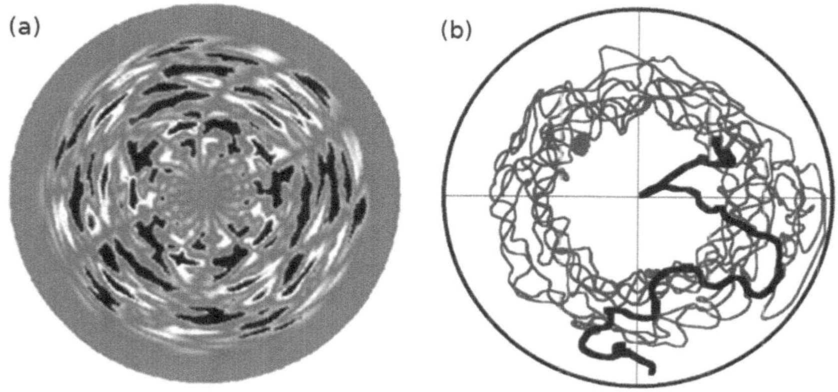

Fig. 6.5 Three-dimensional turbulence model driven by pressure [140]: **a** Fluctuating electrostatic potential numerically computed at a specific time (the observed eddies contribute to trapping tracer particles, as depicted in panel b, resulting in anomalous diffusion); **b** Particle tracers trapped by turbulent eddies (as illustrated in panel a) and undergoing substantial displacements due to avalanche-like events. *Source* [140]

deform over time, characteristic of a viscous fluid. Moreover, it has some degree of elasticity. Hence, blood flows through the blood vessels thanks to its viscosity. Its elasticity allows it to revert to its original shape to some extent after deformation, indicating a certain level of memory. This viscous and elastic combination bestows blood a viscoelastic nature [145, 146]. Understanding the viscoelastic behavior of blood is vital in a variety of medical contexts [147], for instance, (i) cardiovascular diseases (atherosclerosis, hemodynamic disorders), (ii) respiratory diseases, (iii) cancer (tumor growth and angiogenesis), (iv) endocrine disorders, (v) neurological disorders, (vi) glucose-insulin dynamics, and (vii) the design of medical devices like stents and catheters, etc. Moreover, viscoelasticity is also relevant in hemorheology (the study of blood flow and its behavior), which helps in the development of treatments for conditions like thrombosis and atherosclerosis [148, 149].

FDs are helpful in modeling non-Newtonian and viscoelastic behaviors in mathematical models. In the context of blood flow, FDs can be employed for a more precise modeling and analysis of blood behavior within vessels. Blood flow in small vessels and capillaries can exhibit anomalous transport behavior; FDs are especially useful in describing such behavior [150]. Given that blood vessels have complex fractal-like geometries [151], FDs are a natural framework for describing and analyzing systems with fractal properties [152, 153]. They provide a more nuanced and precise depiction of blood behavior under different physiological and pathological conditions, offering the potential for improved medical treatments and interventions.

Models based on FDs are particularly helpful in understanding atherosclerosis because they allow more accurate modeling of the complex blood flow dynamics within narrowed or blocked arteries. Atherosclerosis is marked by the accumulation of fatty deposits (plaque) on the walls of arteries, resulting in the narrowing of these vessels, see Fig. 6.6. Such constriction can hinder the normal flow of blood, escalating the possibility of consequential cardiovascular events like heart attacks and strokes. Atherosclerosis is closely associated with factors such as hypertension, diabetes, elevated cholesterol levels, smoking, oxidative stress, and advancing age [154, 155].

In the exploration of coronary atheroma plaque, [156] employed mathematical modeling techniques, utilizing the Navier–Stokes equations, Darcy's law and convection–diffusion–reaction equations for mass balance modeling in the lumen and intima, and the Kedem–Katchalsky equations for the membranes interface, i.e., endothelium. To implement the model, they used an axisymmetric geometrical coronary artery model as an example. Their finding qualitatively captures the atheroma plaque appearing in the intima layer and matches well with experiments. Since their model dealt only with simple geometry, it cannot capture flows in complicated geometries (such as coronary or carotid arteries with bifurcations) and also other peculiar geometries. Choosing FD models allows us to incorporate complex fractal-like geometries, memory and slow diffusion in the model. Many recent studies employed FDs to analyze atherosclerosis and stenosed arteries.

Using FD-based models [157], analysed the non-Newtonian magnetic blood flow in an inclined stenosed artery. They utilized the Casson fluid to simulate blood flow driven by a uniformly distributed magnetic field and an oscillating pressure gradient. The governing FD equations were formulated using the Caputo-Fabrizio fractional derivative without a singular kernel. In the study of [158], a variable fractional order coronary artery model was proposed. They investigated chaotic dynamics associated with the system under different external disturbances. They found that the inclusion of noise and non-constant derivative order functions introduces complexity to the system, consequently elevating the risk of vascular disease. Yadeta and Shaw [159] reported unsteady non-Newtonian Jeffrey blood flow and transport of magnetic nanoparticles through an inclined porous artery with stenosis using the FD. Concurrently, a novel mathematical modeling of an atherosclerosis initiation, incorporating the Atangana–Baleanu fractional derivative with a memory function, was proposed by [160]. These investigations have contributed to the fundamental understanding of atherosclerosis and stenosed arteries. Recent advancements have

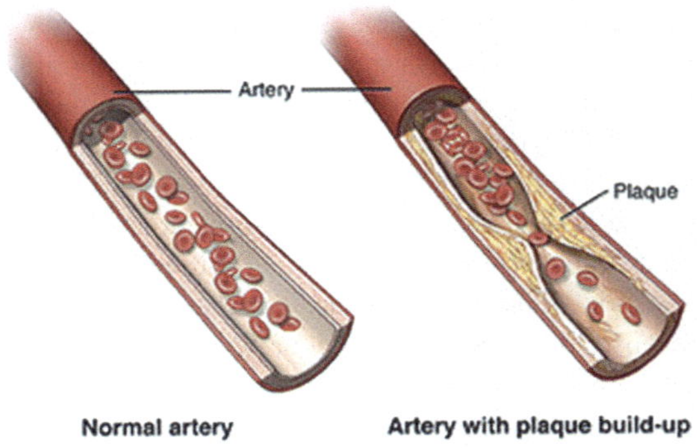

Fig. 6.6 Thickening or hardening of the arteries is referred to as atheroscleros. It is caused by a buildup of plaque in the inner part of the artery. *Source* hopkinsmedicine.org

expanded on these insights by incorporating the viscoelasticity of blood flow in models, providing analyses of atherosclerosis and carotid plaque, as exemplified in [161, 162]. To put it concisely, utilizing viscoelastic flow models based on FD is a promising avenue for enhancing our understanding of blood flow dynamics in conditions like atherosclerosis, stenosed arteries, carotid plaque, and other related health issues. This approach may offer valuable insights for addressing future challenges, with the previously discussed methods seamlessly adaptable.

6.3.9 Active Particles in Viscoelastic Fluid

Active matter—a collection of self-driven particles—is an enthralling subject of study in both the natural and artificial worlds [163]. Microorganisms such as bacteria, algae, and specific cells have innate processes that allow them to move autonomously in natural environments. These self-propelled creatures navigate their environs by responding to numerous cues and frequently engage in collective activities such as swarming or flocking. Scientists are constructing artificial active particles, ranging from nanoscale robots to colloidal particles, that can move in response to planned propulsion mechanisms or chemical reactions inbuilt systems. This field has far-reaching consequences for many areas. For example, understanding the behavior of active particles can light up cellular processes and the dynamics of biological systems in biophysics [164, 165]. It impacts the design of colloidal systems in materials science and can lead to improved medication delivery and targeted therapy. Furthermore, studying active particles guides the construction of autonomous agents capable of navigating complex environments in robotics. Researchers are uncovering new pathways for technological growth and getting more profound insights into the behavior of live beings at the microscale by examining the underlying principles governing the mobility and interactions of these dynamic entities.

The interaction of active particles in viscoelastic fluids is a fascinating junction of physics and materials science. When introduced into viscoelastic fluids, active particles navigate the viscoelastic medium in a way that is impacted by both their self-propelled behavior and the fluid's intricate rheological features. This interaction frequently results in intriguing phenomena such as sub-diffusion. Understanding the behavior of active particles in viscoelastic fluids offers great potential for a variety of applications ranging from targeted medication delivery in biological systems to the design of new materials with tailored features. Furthermore, it offers a glimpse into the subtle dynamics that occur at the microscale inside complex fluid environments. McKinley et al. [166] undertook the mathematical modeling of transient anomalous diffusion of tracer particles within viscoelastic fluids. Employing the generalized Langevin equation, which serves as the standard model for diffusion in viscoelastic fluids, they derived closed-form solutions. Their models extend beyond the traditional Rouse model in terms of the scaling properties of the diffusive memory time scale spectrum. They demonstrated that control over the time window of sub-diffusion and the transient sub-diffusive MSD power law exponent, encompassing the entire sub-diffusive scaling range, can be obtained directly from the parameterized memory spectrum.

Similarly, sub-diffusive behavior has also been observed in active particles in the viscoelastic medium, e.g., active Brownian particles in the viscoelastic medium. Active Brownian particle (ABP) is a model of self-propelled motion in a dissipative environment. When ABPs are introduced into viscoelastic environments, their behavior becomes influenced by the rheological properties of the medium. This interplay often leads to sub-diffusion. Precisely characterizing the dynamics of an ABP as it interacts with a viscoelastic polymer environment poses a significant scientific challenge. The investigation conducted by [167] revealed that the ABP cross-linker exhibits a noticeable anomalous diffusion pattern within the Rouse regime, influenced by viscoelastic feedback. As the Péclet number increases, the system attunes a smaller exponent, resulting in greater sub-diffusivity. This implies that a particle with heightened activity demonstrates a more pronounced sub-diffusive behavior when measuring the MSD. Sub-diffusive behavior can also be seen in confined geometries. For instance, sub-diffusion in ABP has been reported under lateral parabolic confinement by [168]. They found that motion is sub-diffusion to ballistic depending on angular noise strength and confinement intensity. They discovered that sub-diffusion happens when particles do not have much space to move around. This can be because the area is minimal or because of a lot of background "noise" in the environment. Super-diffusion has been reported even at very high confinement. Owing to diffusion, clusters exist for given values of confinement and angular noise.

Crowded systems, i.e., "living systems at the sub-cellular, cellular, and multicellular levels" are another type of active matter that exhibits sub-diffusion [169–172]. Examples of active crowded molecular systems include the cell cytoplasm, colonies of bacteria, biofilms, tissues, enzymes in cellular reactions, synthetic nanoparticles in biological systems, etc. The complex behavior of surrounding particles impacts all individual particles within a crowded environment [172]. Because they absorb and release energy locally, active particles are frequently found in crowded environments where they are constantly in motion. At the molecular level, these active particles may function as tiny engines, moving the cytoplasm and dragging objects around inside cells. Another possibility is that they are unusually fast-moving enzymes. Larger cellular scales may contain self-moving cells that swim in a liquid or make their way through tissues or collections of cells. Because of this activity, equilibrium statistical mechanics cannot describe these systems. Researchers have employed both continuum models and lattice models to address this issue. They have observed an anomalous diffusion for a tracer in crowded active systems, both in models that consider hydrodynamic interactions and those that do not. Abbaspour and Klumpp [172] studied a simple lattice-based model. In this model, particles could collide with each other, and different types of particles could move at different speeds with varying coefficients of diffusion. They [172] used this model to see how the movement of a tracer particle is affected by how active the particles are and how many of them there are around, without considering how fluids might influence the motion. They did not consider hydrodynamic interactions in their model. They also showed that when active particles group together, it affects how the tracer particle moves, slowing down the enhanced diffusion effect.

To what follows, understanding exactly how an active Brownian particle (ABP) moves when it interacts with a stretchy polymer environment, confined geometries, and/or background noise is a tough scientific puzzle to solve. Similarly, many simulations and experiments show complex dynamics and sub-diffusion in active, crowded systems such as the cytoplasm of cells. There are several experimental and simulation studies, which showed sub-diffusion in ABP motion as well as in active crowded systems. However, rigorous mathematical theory is far from understanding. Comprehending the motion of active particles is paramount for gaining insights into diverse biological processes, including intracellular transport, cellular migration, and tissue dynamics. This understanding is especially crucial for advancing fields, such as medicine, where precise control over particle motion is imperative. Incorporating FDs into diffusion equations enhances these models by accommodating non-local and memory-dependent effects. This method more accurately depicts the intricate dynamics of active particles in viscoelastic media. Consequently, researchers can quantitatively analyze and predict particle behavior in complex biological systems, effectively bridging the theoretical understanding with practical applications in biomedicine and beyond. The methodologies outlined in this book offer an adaptable framework that can be readily extended to analyze some physical aspects of active particles.

6.3.10 Microfluidic Devices

Microfluidic devices have gained significance for their role in micro-electro-mechanical systems and their applications in microbiological sensors, exemplified by lab-on-a-chip technologies [173]. Microfluidic devices consist of channels, chambers, and valves fabricated on a small chip, usually made of glass, silicon, or polymers. These channels are typically on the order of tens to hundreds of micrometers in width. The small dimensions of these channels allow for precise control over fluid flow. Such devices play a vital role in various fields, including bioengineering (such as DNA analysis), as well as in facilitating chemical transport and managing heat transfer within electronic components [174].

At the core of microfluidic devices lies the interplay between viscoelastic fluids and external electric fields at scales as minute as the micro- and nano-levels. These interactions fall into two classes: electrohydrodynamics (EHD) and electrokinetic instabilities (EKI), collectively referred to as electro-elastic instabilities (EEI), see Fig. 6.7. In EHD, a distinctive feature is the accumulation of a net charge at the boundaries. Conversely, in EKI, what sets it apart is the creation of electrically neutral double layers. These fundamental distinctions shape the behavior and phenomena observed in these systems, ultimately driving advances in fields ranging from biotechnology to materials science to electronic components. EEI can induce alterations in viscosity and shear-thinning behavior, prompt spontaneous deformations or

Fig. 6.7 Classification of electro-elastic instabilities

buckling of structures, and generate complex fluid flow patterns, all of which can be strategically utilized for a wide range of applications. It is worth noticing that efficient mixing is essential to enhance the performance of miniature devices, especially in clinical setups, which rely on small samples where fast and complete reactions are needed. Mixing is usually difficult in microchannel flows, because at low Reynolds number, the flow is typically laminar, and the process is limited by molecular diffusion. Thus, the EEI flow forcing methods are particularly useful in inducing flow instabilities and are suitable for the miniaturization process—which involves reducing the size of components, structures, or entire systems without compromising their performance or capabilities.

Studies on EEIs of classical viscoelastic fluids are extensive. Some notable studies are the flow past parallel plate channels [175] and microtubes [176], where Maxwell and Oldroyd-B models were considered, respectively. Numerous previous experimental and numerical investigations have shown that the EKI phenomenon induces a chaotic flow field within microdevices for Newtonian fluids, thereby enhancing fluid mixing. In a recent study by [177], the EKI phenomenon was numerically investigated in a microfluidic T junction, where two viscoelastic fluids with different electrical conductivities were transported side by side under the influence of an electric field. The findings indicated that chaotic convection resulting from EKI can be mitigated when dealing with viscoelastic fluids in contrast to Newtonian ones. By carefully adjusting the Weissenberg number and polymer viscosity ratio, the chaotic fluctuations within a T microfluidic junction can be reduced. Interestingly, this study demonstrated that suppressing chaotic fluctuations, especially in the spanwise direction, hinders the mixing of the two viscoelastic fluids. Consequently, careful consideration is essential when dealing with mixing viscoelastic fluids in the presence of EKI.

In recent times, FD-based models have been employed to investigate various aspects of EEI, such as abnormal diffusion, slip flow dynamics, and the formation of electrically neutral double layers in viscoelastic flows within microchannels. These models provide insights into phenomena that would be challenging to comprehend using traditional differential equations alone. For instance, [178] utilized the temporal Caputo FD [179] to examine the EHD of generalized Maxwell fluids under rotating slip flow driven by a periodic electric field. The fractional second-order fluid in a microchannel was reported in Ref. [180]. A numerical study of the EHD of Maxwell fluids based on RL-FDs in a microchannel was done in Ref. [181]. They presented a conjugate gradient method to solve their model by keeping nonlinear terms. The fractional-order Oldroyd-B flow between two parallel plates was studied by [182] where they focused on heat flow.

Studies [178, 180–182] showed that models incorporating FDs play a pivotal role in the analysis of EEI in microfluidic devices due to their ability to capture the non-local, complex nature of the underlying phenomena. In the microfluidics realm, where fluid behaviors are constrained by small-scale geometries, incorporating FDs in mathematical models enables a more accurate representation of the viscoelastic properties of fluids and their response to applied electric fields. This is particularly crucial for understanding EEIs, encompassing EHD and EKI, as these phenomena, involve complex interactions between fluids and electric fields at the micro- and nanoscales. These models contribute significantly to knowledge advancement by enabling more precise predictions and optimizations for a wide range of applications ranging from biotechnology to materials science which points towards a promising future research domain.

References

1. Koeller RC (1984) Applications of fractional calculus to the theory of viscoelasticity. J Appl Mech 51(2):299–307. https://doi.org/10.1115/1.3167616
2. Liu Y, Yu D, Zhao H, Wen J, Wen X (2008) Theoretical study of two-dimensional phononic crystals with viscoelasticity based on fractional derivative models. J Phys D: Appl Phys 41(6):065–503. https://doi.org/10.1088/0022-3727/41/6/065503
3. Ferr'as L, Ford NJ, Morgado ML, Rebelo M, McKinley GH, Nóbrega JM (2018) Theoretical and numerical analysis of unsteady fractional viscoelastic flows in simple geometries. Comput Fluids 174:14–33. http://dx.doi.org/j.compfluid.2018.07.004
4. Heymans N, Bauwens J (1994) Fractal rheological models and fractional differential equations for viscoelastic behavior. Rheola Acta 33:210–219. https://doi.org/10.1007/BF00437306
5. Schiessel H, Metzler R, Blumen A, Nonnenmacher TF (1995) Generalized viscoelastic models: their fractional equations with solutions. J Phys A: Math Gen 28:23. https://doi.org/10.1088/0305-4470/28/23/012
6. Hoffmann KH, Prehl J (2008) Anomalous transport on disordered fractals. In: Rainer Klages GR, Sokolov IM (eds.) Anomalous transport: foundations and applications. Wiley, Dordrecht, the Netherlands, pp 397–427
7. Failla G, Zingales M (2020) Advanced materials modelling via fractional calculus: challenges and perspectives. Philos Trans Royal Soc A 378(20200):050. https://doi.org/10.1098/rsta.2020.0050
8. Nigmatullin R (1992) Fractional integral and its physical interpretation. Theor Math Phys 90:242–251. https://doi.org/10.1007/BF01036529
9. Rutman RS (1995) On physical interpretations of fractional integration and differentiation. Theor Math Phys 105:1509–1519. https://doi.org/10.1007/BF02070871
10. Sabatier J, Farges C, Trigeassou J-C (2014) Fractional systems state space description: some wrong ideas and proposed solutions. J Vib Control 20:1076–1084. https://doi.org/10.1177/1077546313481839
11. Sabatier J, Farges C, Merveillaut M, Feneteau L (2012) On observability and pseudo state estimation of fractional order systems. Eur J Control 18:260–271. https://doi.org/10.3166/ejc.18.260-271

12. Lazopoulos K (2006) Non-local continuum mechanics and fractional calculus. Mech Res Commun 33:753–757. https://doi.org/10.1016/j.mechrescom.2006.05.001
13. Sabatier J, Farges C (2018) Comments on the description and initialization of fractional partial differential equations using Riemann-Liouville's and Caputo's definitions. J Comput Appl Math 339:30–39. https://doi.org/10.1016/j.cam.2018.02.030
14. Yang X-J (2019) General fractional derivatives: theory, methods and applications. CRC Press
15. Mainardi F (2020) Why the Mittag-Leffler function can be considered the queen function of the fractional calculus? Entropy (Basel) 22:1359. https://doi.org/10.3390/e22121359
16. Kiryakova V (2010) The multi-index mittag-leffler functions as an important class of special functions of fractional calculus. Comput Math Appl 59:1885–1895. https://doi.org/10.1016/j.camwa.2009.08.025
17. Baleanu D, Diethelm K, Scalas E, Trujillo JJ (2012) Fractional calculus: models and numerical methods, vol 3. World Scientific
18. Podlubny I (2001) Geometric and physical interpretation of fractional integration and fractional differentiation. arXiv: Classical Analysis and ODEs
19. Tarasov VE (2016) Geometric interpretation of fractional-order derivative. Fract Calc Appl Anal 19(5):1200–1221. https://doi.org/10.1515/fca-2016-0062
20. de Oliveira EC, Tenreiro Machado JA (2014) A review of definitions for fractional derivatives and integral. Math Probl Eng 238:459. https://doi.org/10.1155/2014/238459
21. Valério D, Ortigueira MD, Lopes AM (2022) How many fractional derivatives are there? Mathematics 10(5). https://doi.org/10.3390/math10050737
22. Oldham KB, Spanier J (1974) The fractional calculus: integrations and differentiations of arbitrary order. WAcademic Press, New York
23. Miller KS, Ross B (2013) An introduction to the fractional calculus and fractional Differential Equations. Wiley, New York
24. De Oliveira EC, Tenreiro Machado JA et al (2014) A review of definitions for fractional derivatives and integral. Math Probl Eng 2014:238459. https://doi.org/10.1155/2014/238459
25. Atangana A (2018) Chapter 5 - fractional operators and their applications. In: Atangana A (ed) Fractional operators with constant and variable order with application to geo-hydrology. Academic Press, pp 79–112. 10.1016/B978-0-12-809670-3.00005-9
26. Caputo M, Fabrizio M (2015) A new definition of fractional derivative without singular kernel. Prog Fract Differ Appl 1:73–85. https://doi.org/10.12785/pfda/010201
27. Sabatier J (2020) Non-singular kernels for modelling power law type long memory behaviours and beyond. Cybern Syst 51:383–401. https://doi.org/10.1080/01969722.2020.1758470
28. Leygue A, Bailly C, Keunings R (2006) A tube-based constitutive equation for polydisperse entangled linear polymers. J Non-Newt Fluid Mech
29. Uchaikin VV (2013) Fractional derivatives for physicists and engineers, vol 2. Springer
30. Sun H, Zhang Y, Baleanu D, Chen W, Chen Y (2018) A new collection of real world applications of fractional calculus in science and engineering. Commun Nonlinear Sci Numer Simul 64:213–231. https://doi.org/10.1016/j.cnsns.2018.04.019
31. Hilfer R (2000) Applications of fractional calculus in physics. World scientific
32. Song L, Yu W, Tan Y, Duan K (2024) Calculations of fractional derivative option pricing models based on neural network. J Comput Appl Math 437(115):462. https://doi.org/10.1016/j.cam.2023.115462
33. Chen W, Sun H, Li X (2023) Fractional derivative modeling in mechanics and engineering. Springer, Singapore
34. Petráš I (2011) Fractional-order nonlinear systems: modeling, analysis and simulation. Springer Science & Business Media
35. Yavuz M, Özdemir N (eds) (2022) Fractional calculus: new applications in understanding nonlinear phenomena. Bentham Science Publishers
36. Ambaliya M, Bera A (2023) A perspective review on the current status and development of polymer flooding in enhanced oil recovery using polymeric nanofluids. Ind Eng Chem Res 62:2444–2459. https://doi.org/10.1021/acs.iecr.2c04582

37. Alvarado V, Manrique E (2010) Enhanced oil recovery: field planning and development strategies. Gulf Professional Publishing
38. Mohsenatabar Firozjaii A, Saghafi HR (2020) Review on chemical enhanced oil recovery using polymer flooding: fundamentals, experimental and numerical simulation. Petroleum 6(2):115–122. https://doi.org/10.1016/j.petlm.2019.09.003
39. Al-Hajri S, Mahmood SM, Abdulrahman A, Abdulelah H, Akbari S, Saraih N (2019) An experimental study on hydrodynamic retention of low and high molecular weight sulfonated polyacrylamide polymer. Polymers 11(9). https://doi.org/10.3390/polym11091453
40. Speight JG (2013) Heavy oil production processes. Gulf Professional Publishing
41. Dake L (1978) Fundamentals of petroleum engineering. Elsivier
42. Delamaide E, Tabary R, Rousseau D (2014) Chemical EOR in low permeability reservoirs. In: SPE EOR conference at oil and gas West Asia. OnePetro
43. Marliere C, Wartenberg N, Fleury M, Tabary R, Dalmazzone C, Delamaide E (2015) Oil recovery in low permeability sandstone reservoirs using surfactant-polymer flooding. In: SPE Latin America and Caribbean petroleum engineering conference, p D031S030R004. SPE
44. Guetni I, Marlière C, Rousseau D, Pelletier M, Bihannic I, Villiéras F (2020) Transport of eor polymer solutions in low permeability porous media: impact of clay type and injection water composition. J Pet Sci Eng 186(106):690. https://doi.org/10.1016/j.petrol.2019.106690
45. Sorbie KS (2013) Polymer-improved oil recovery. Springer
46. Berg CF, Danilova A, Liu M (2019) Upscaling of polymer adsorption. J Pet Sci Eng 179:655–668. https://doi.org/10.1016/j.petrol.2019.03.079
47. Beteta A, Sorbie KS, Skauge A (2022) Immiscible viscous fingering: the simulation of tertiary polymer displacements of viscous oils in 2d slab floods. Polymers 14(19). https://doi.org/10.3390/polym14194159
48. Ahmadikhamsi S, Golfier F, Oltean C, Lefèvre E, Bahrani SA (2020) Impact of surfactant addition on non-Newtonian fluid behavior during viscous fingering in Hele-Shaw cell. Phys Fluids 32(1):012–103. https://doi.org/10.1063/1.5128589
49. Beteta A, Sorbie KS, Johnson G (2023) Immiscible viscous fingering at the field scale: numerical simulation of the captain polymer flood. SPE J 1–15. https://doi.org/10.2118/215813-PA
50. Shokri H, Kayhani MH, Norouzi M (2017) Nonlinear simulation and linear stability analysis of viscous fingering instability of viscoelastic liquids. Phys Fluids 29(3):033–101. https://doi.org/10.1063/1.4977443
51. Ochoa-Tapia JA, Valdes-Parada FJ, Alvarez-Ramirez J (2007) A fractional-order darcy's law. Phys A Stat Mech Appl 374:1–14
52. Zhou H, Yang S (2018) Fractional derivative approach to non-darcian flow in porous media. J Hydrol 566:910–918. https://doi.org/10.1016/j.jhydrol.2018.09.039
53. Zhou H, Yang S, Zhang S (2019) Modeling non-darcian flow and solute transport in porous media with the caputo-fabrizio derivative. Appl Math Model 68:603–615. https://doi.org/10.1016/j.apm.2018.09.042
54. Deseri L, Zingales M (2015) A mechanical picture of fractional-order darcy equation. Commun Nonlinear Sci Numer Simul 20:940–949. https://doi.org/10.1016/j.cnsns.2014.06.021
55. Das SK, Choi SU, Yu W, Pradeep T (2007) Nanofluids: science and technology. Wiley
56. Volgin IV, Larin SV, Lyulin SV (2018) Diffusion of nanoparticles in polymer systems. Polym Sci Ser C 60:122–134. https://doi.org/10.1134/S1811238218020212
57. Guo H, Bourret G, Lennox RB, Sutton M, Harden JL, Leheny RL (2012) Entanglement-controlled subdiffusion of nanoparticles within concentrated polymer solutions. Phys Rev Lett 109(055):901. https://doi.org/10.1103/PhysRevLett.109.055901
58. Aramideh S, Vlachos PP, Ardekani AM (2019) Nanoparticle dispersion in porous media in viscoelastic polymer solutions. J Non-Newton Fluid Mech 268:75–80. https://doi.org/10.1016/j.jnnfm.2019.05.003
59. Cao Z, Zhao J, Wang Z, Liu F, Zheng L (2016) Mhd flow and heat transfer of fractional maxwell viscoelastic nanofluid over a moving plate. J Mol Liq 222:1121–1127. https://doi.org/10.1016/j.molliq.2016.08.012

60. Liu L, Feng L, Xu Q, Zheng L, Liu F (2020) Flow and heat transfer of generalized maxwell fluid over a moving plate with distributed order time fractional constitutive models. Int Commun Heat Mass Transf 116(104):679. https://doi.org/10.1016/j.icheatmasstransfer.2020.104679
61. Li C, Zeng F (2015) Numerical methods for fractional calculus. Chapman and Hall/CRC
62. Umbanhowar PB, Melo F, Swinney HL (1996) Localized excitations in a vertically vibrated granular layer. Nature 382:793–796. https://doi.org/10.1038/382793a0
63. Daerr A, Douady S (1999) Two types of avalanche behaviour in granular media. Nature 399:241–243. https://doi.org/10.1038/20392
64. Goldhirsch I (2003) Rapid granular flows. Annu Rev Fluid Mech 35:267–293. https://doi.org/10.1146/annurev.fluid.35.101101.161114
65. Marty G, Dauchot O (2005) Subdiffusion and cage effect in a sheared granular material. Phys Rev Lett 94(015):701. https://doi.org/10.1103/PhysRevLett.94.015701
66. Reis PM, Ingale RA, Shattuck MD (2007) Caging dynamics in a granular fluid. Phys Rev Lett 98(188):301. https://doi.org/10.1103/PhysRevLett.98.188301
67. Fullmer WD, Hrenya CM (2017) The clustering instability in rapid granular and gas-solid flows. Annu Rev Fluid Mech 49:485–510. https://doi.org/10.1146/annurev-fluid-010816-060028
68. Gray JMNT (2018) Particle segregation in dense granular flows. Annu Rev Fluid Mech 50:407–433. https://doi.org/10.1146/annurev-fluid-122316-045201
69. Zhou GGD, Cui KFE, Jing L, Zhao T, Song D, Huang Y (2020) Particle size segregation in granular mass flows with different ambient fluids. J Geophys Res Solid Earth 125:e2020JB019 536. https://doi.org/10.1029/2020JB019536
70. Ottino JM, Khakhar DV (2000) Mixing and segregation of granular materials. Annu Rev Fluid Mech 32(1):55–91. https://doi.org/10.1146/annurev.fluid.32.1.55
71. Tegzes P, Vicsek T, Schiffer P (2002) Avalanche dynamics in wet granular materials. Phys Rev Lett 89(094):301. https://doi.org/10.1103/PhysRevLett.89.094301
72. Zuriguel I, Gray J, Peixinho J, Mullin T (2006) Pattern selection by a granular wave in a rotating drum. Phys Rev E 73(061):302. https://doi.org/10.1103/PhysRevE.73.061302
73. Christov IC, Stone HA (2012) Resolving a paradox of anomalous scalings in the diffusion of granular materials. Proc Natl Acad Sci 109(40):16 012–16 017. https://doi.org/10.1073/pnas.1211110109
74. Bodrova A, Dubey AK, Puri S, Brilliantov N (2012) Intermediate regimes in granular brownian motion: Superdiffusion and subdiffusion. Phys Rev Lett 109(178):001. https://doi.org/10.1103/PhysRevLett.109.178001
75. Scalliet C, Gnoli A, Puglisi A, Vulpiani A (2015) Cages and anomalous diffusion in vibrated dense granular media. Phys Rev Lett 114(198):001. https://doi.org/10.1103/PhysRevLett.114.198001
76. Utter B, Behringer RP (2004) Self-diffusion in dense granular shear flows. Phys Rev E 69(031):308. https://doi.org/10.1103/PhysRevE.69.031308
77. Cone RA (2009) Barrier properties of mucus. Adv Drug Deliv Rev 61:75–85. https://doi.org/10.1016/j.addr.2008.09.008
78. Witten J, Samad T, Ribbeck K (2018) Selective permeability of mucus barriers. Curr Opin Biotechnol 52:124–133. https://doi.org/10.1016/j.copbio.2018.03.010
79. Ernst M, John T, Guenther M, Wagner C, Schaefer UF, Lehr CM (2017) A model for the transient subdiffusive behavior of particles in mucus. Biophys J 112(1):172–179. https://doi.org/10.1016/j.bpj.2016.11.900
80. Kaunitz JD (1999) Barrier function of gastric mucus. Keio J Med 48:63–68. https://doi.org/10.2302/kjm.48.63
81. Lewis OL, Keener JP (2021) Enhanced electrodiffusive transport across a mucus layer. SIAM J Appl Math 81:965–981. https://doi.org/10.1137/20M1348327
82. Bahari HM, Ross IN, Turnberg LA (1982) Demonstration of a ph gradient across the mucus layer on the surface of human gastric mucosa in vitro. Gut 23(6):513–516. https://doi.org/10.1136/gut.23.6.513

83. Newby JM, Seim I, Lysy M, Ling Y, Huckaby J, Lai SK, Forest MG (2018) Technological strategies to estimate and control diffusive passage times through the mucus barrier in mucosal drug delivery. Adv Drug Deliv Rev 124:64–81. https://doi.org/10.1016/j.addr.2017.12.002
84. Barr JJ, Auro R, Sam-Soon N, Kassegne S, Peters G, Bonilla N, Hatay M, Mourtada S, Bailey B et al (2015) Subdiffusive motion of bacteriophage in mucosal surfaces increases the frequency of bacterial encounters. Proc Natl Acad Sci 112:13 675–13 680. https://doi.org/10.1073/pnas.1508355112
85. Heatley N (1959) Mucosubstance as a barrier to diffusion. Gastroenterology 37(3):313–317. https://doi.org/10.1016/S0016-5085(19)35420-4
86. Livingston EH, Engel E (1995) Modeling of the gastric gel mucus layer: application to the measured ph gradient. J Clin Gastroenterol 21(suppl. 1):S120–S124. http://dx.doi.org/https://pubmed.ncbi.nlm.nih.gov/8775003/
87. Pfeiffer CJ (1981) Experimental analysis of hydrogen ion diffusion in gastrointestinal mucus glycoprotein. Am J Physiol - Gastrointest Liver Physiol 240(2):G176–G182. PMID: 7468808 https://doi.org/10.1152/ajpgi.1981.240.2.G176
88. Williams S, Turnberg L (1980) Retardation of acid diffusion by pig gastric mucus: a potential role in mucosal protection. Gastroenterology 79(2):299–304. https://doi.org/10.1016/0016-5085(80)90145-6
89. Marczynski M, Käsdorf BT, Altaner B, Wenzler A, Gerland U, Lieleg O (2018) Transient binding promotes molecule penetration into mucin hydrogels by enhancing molecular partitioning. Biomater Sci 6:3373–3387. https://doi.org/10.1039/C8BM00664D
90. Lewis OL, Keener JP (2021) Enhanced electrodiffusive transport across a mucus layer. SIAM J Appl Math 81(3):965–981. https://doi.org/10.1137/20M1348327
91. Lewis OL, Missey E, Keener JP (2021) Principles of slowed hydrogen diffusion through a mucus layer. Phys Rev E 104(044):403. https://doi.org/10.1103/PhysRevE.104.044403
92. Ernst M, John T, Guenther M, Wagner C, Schaefer UF, Lehr C-M (2017) A model for the transient subdiffusive behavior of particles in mucus. Biophys J 112(1):172–179. https://doi.org/10.1016/j.bpj.2016.11.900
93. Jenkins JT (1978) Flows of nematic liquid crystals. Annu Rev Fluid Mech 10(1):197–219. https://doi.org/10.1146/annurev.fl.10.010178.001213
94. Chandrasekhar S, Madhusudana NV (1980) Liquid crystals. Annu Rev Mater Sci 10(1):133–155. https://doi.org/10.1146/annurev.ms.10.080180.001025
95. Andrienko D (2018) Introduction to liquid crystals. J Mol Liq 267:520–541. https://doi.org/10.1016/j.molliq.2018.01.175
96. Ericksen JL (1961) Conservation laws for liquid crystals. Trans Soc Rheol 5(1):23–34. https://doi.org/10.1122/1.548883
97. Leslie FM (1968) Some constitutive equations for liquid crystals. Arch Ration Mech Anal 28:265–283. https://doi.org/10.1007/BF00251810
98. Ericksen J (1968) Twisting of partially oriented liquid crystals. Q Appl Math 25(4):474–479
99. Oseen CW (1929) Die anisotropen flüssigkeiten tatsachen und theorien. Physik Physik Chem 20:25–113
100. Frank FC (1958) I. liquid crystals on the theory of liquid crystals. Discuss Faraday Soc 25:19–28
101. Sánchez-Miranda M, Bonilla-Capilla B, Sarmiento-Gómez E, Lázaro-Lázaro E, Ramírez-Saito A, Medina-Noyola M, Arauz-Lara J (2015) Transition from diffusive to subdiffusive motion in colloidal liquids. Soft Matter 11:655–658. https://doi.org/10.1039/C4SM02078B
102. De Gaetani L, Prampolini G, Tani A (2008) Subdiffusive dynamics of a liquid crystal in the isotropic phase. J Chem Phys 128(194):501. https://doi.org/10.1063/1.2916681
103. Maeda K, Narumi T, Anugraha R, Okabe H, Hara K, Hidaka Y (2018) Sub-diffusion in electroconvective turbulence of homeotropic nematic liquid crystals. J Phys Soc Jpn 87(014):401. https://doi.org/10.7566/JPSJ.87.014401
104. Coscia BJ, Shirts MR (2020) Capturing subdiffusive solute dynamics and predicting selectivity in nanoscale pores with time series modeling. J Chem Theory Comput 16(9):5456–5473. https://doi.org/10.1021/acs.jctc.0c00445

105. Williams R (2004) Domains in liquid crystals. J Chem Phys 39:384–388. https://doi.org/10.1063/1.1734257
106. Brinkman WF, Cladis PE (1982) Defects in liquid crystals. Phys Today 35(5):48–54. https://doi.org/10.1063/1.2915094
107. Kleman M (1989) Defects in liquid crystals. Rep Prog Phys 52:555. https://doi.org/10.1088/0034-4885/52/5/002
108. Turiv T, Lazo I, Brodin A, Lev BI, Reiffenrath V, Nazarenko VG, Lavrentovich OD (2013) Effect of collective molecular reorientations on brownian motion of colloids in nematic liquid crystal. Science 342:1351–1354. https://doi.org/10.1126/science.1240591
109. Pagès JM, Ignés-Mullol J, Sagués F (2019) Anomalous diffusion of motile colloids dispersed in liquid crystals. Phys Rev Lett 122(198):001. https://doi.org/10.1103/PhysRevLett.122.198001
110. Kramer L, Pesch W (1995) Convection instabilities in nematic liquid crystals. Annu Rev Fluid Mech 27:515–539. https://doi.org/10.1146/annurev.fl.27.010195.002503
111. Lavrentovich OD (2014) Transport of particles in liquid crystals. Soft Matter 10:1264–1283. https://doi.org/10.1039/C3SM51628H
112. Orihara H, Fukase A, Ishibashi Y (1995) Nonlinear dielectric spectroscopy of the goldstone mode in a ferroelectric liquid crystal. J Phys Soc Jpn 64(3):976–980. https://doi.org/10.1143/JPSJ.64.976
113. Gómez-Aguilar J, Atangana A (2017) Fractional hunter-saxton equation involving partial operators with bi-order in riemann-liouville and liouville-caputo sense. Eur Phys J Plus 132:100. https://doi.org/10.1140/epjp/i2017-11371-6
114. Ciuchi F, Mazzulla A, Scaramuzza N, Lenzi EK, Evangelista LR (2012) Fractional diffusion equation and the electrical impedance: Experimental evidence in liquid-crystalline cells. J Phys Chem C 116(15):8773–8777. https://doi.org/10.1021/jp211097m
115. Ribeiro de Almeida R, Evangelista L, Lenzi E, Zola R, Jákli A (2019) Electrical transport properties and fractional dynamics of twist-bend nematic liquid crystal phase. Commun Nonlinear Sci Numer Simul 70:248–256. https://doi.org/10.1016/j.cnsns.2018.10.021
116. Owolabi KM (2020) Numerical simulation of nonlinear ecological models with nonlocal and nonsingular fractional derivative. Springer, Singapore, pp 303–320. 10.1007/978-981-15-2286-4_10
117. Kumar P, Erturk VS (2021) Environmental persistence influences infection dynamics for a butterfly pathogen via new generalised caputo type fractional derivative. Chaos Solit Fractals 144(110):672. https://doi.org/10.1016/j.chaos.2021.110672
118. Ghanbari B (2021) On the modeling of an eco-epidemiological model using a new fractional operator. Results Phys 21(103):799. https://doi.org/10.1016/j.rinp.2020.103799
119. Hapca S, Crawford JW, Young IM (2009) Anomalous diffusion of heterogeneous populations characterized by normal diffusion at the individual level. J R Soc Interface 6(30):111–122. https://doi.org/10.1098/rsif.2008.0261
120. Pimenov A, Kelly T, Korobeinikov A, O'Callaghan M, Pokrovskii A, Rachinskii D (2012) Memory effects in population dynamicsâŁ¯: Spread of infectious disease as a case study. Math Model Nat Phenom 7(3):204–226. https://doi.org/10.1051/mmnp/20127313
121. Vilk O, Aghion E, Avgar T, Beta C, Nagel O, Sabri A, Sarfati R, Schwartz DK, Weiss M et al (2022) Unravelling the origins of anomalous diffusion: from molecules to migrating storks. Phys Rev Res 4(033):055. https://doi.org/10.1103/PhysRevResearch.4.033055
122. Rietkerk M, Dekker SC, de Ruiter PC, van de Koppel J (2004) Self-organized patchiness and catastrophic shifts in ecosystems. Science 305(5692):1926–1929. https://doi.org/10.1126/science.1101867
123. Reynolds HL, Clay K (2011) Migratory species and ecological processes. Environ Law 41:371–391
124. Lakshmikantham V (1981) Some problems of reaction-diffusion equations. In: de Mottoni P, Salvadori L (eds) Nonlinear differential equations. Academic Press, pp 243–258. 10.1016/B978-0-12-508780-3.50025-0

125. Cosner C (2008) Reaction–diffusion equations and ecological modeling. Springer, Berlin, pp 77–115. 10.1007/978-3-540-74331-6_3
126. Cantrell RS, Cosner C (2004) Spatial heterogeneity in reaction-diffusion models for two competing species. Wiley, New York, pp 295–349. 10.1002/0470871296.ch6
127. Sengupta A, Kruppa T, Löwen H (2011) Chemotactic predator-prey dynamics. Phys Rev E 83(031):914. https://doi.org/10.1103/PhysRevE.83.031914
128. Owolabi KM, Atangana A (2018) Modelling and formation of spatiotemporal patterns of fractional predation system in subdiffusion and superdiffusion scenarios. Eur Phys J Plus 133. https://doi.org/10.1140/epjp/i2018-11886-2
129. Owolabi K, Pindza E (2018) Mathematical and computational studies of fractional reaction-diffusion system modelling predator-prey interactions. J Numer Math 26:97–110. https://doi.org/10.1515/jnma-2016-1044
130. Goldston RJ (2020) Introduction to plasma physics. CRC Press
131. Gurnett DA, Bhattacharjee A (2017) Introduction, 2nd edn. Cambridge University Press, pp 1–3. 10.1017/9781139226059.002
132. Shukla PK, Mamun A (2015) Introduction to dusty plasma physics. CRC Press
133. Wang K, Li W, Huang D, Reichhardt C, Reichhardt CJO, Murillo MS, Feng Y (2018) Structures and diffusion of two-dimensional dusty plasmas on one-dimensional periodic substrates. Phys Rev E 98(063):204. https://doi.org/10.1103/PhysRevE.98.063204
134. Qin G, Matthaeus WH, Bieber JW (2002) Subdiffusive transport of charged particles perpendicular to the large scale magnetic field. Geophys Res Lett 29(4):7–1–7–4. https://doi.org/10.1029/2001GL014035
135. Nunomura S, Samsonov D, Zhdanov S, Morfill G (2006) Self-diffusion in a liquid complex plasma. Phys Rev Lett 96(015):003. https://doi.org/10.1103/PhysRevLett.96.015003
136. Shalchi A (2017) Time-dependent perpendicular transport of energetic particles in magnetic turbulence with transverse complexity. Phys Plasmas 24(050):702. https://doi.org/10.1063/1.4982805
137. Fasoli A, Avino F, Bovet A, Furno I, Gustafson K, Jolliet S, Loizu J, Malinverni D, Ricci P et al (2013) Basic investigations of electrostatic turbulence and its interaction with plasma and suprathermal ions in a simple magnetized toroidal plasma. Nucl Fusion 53(063):013. https://doi.org/10.1088/0029-5515/53/6/063013
138. Anderson J, Moradi S, Rafiq T (2018) Non-linear langevin and fractional fokker-planck equations for anomalous diffusion by lévy stable processes. Entropy 20:760. https://doi.org/10.3390/e20100760
139. Carr HY, Purcell EM (1954) Effects of diffusion on free precession in nuclear magnetic resonance experiments. Phys Rev 94:630–638. https://doi.org/10.1103/PhysRev.94.630
140. del Castillo-Negrete D, Carreras BA, Lynch VE (2004) Fractional diffusion in plasma turbulence. Phys Plasmas 11:3854–3864. https://doi.org/10.1063/1.1767097
141. Kostadinova EG, Banka R, Padgett JL, Liaw CD, Matthews LS, Hyde TW (2021) Fractional Laplacian spectral approach to turbulence in a dusty plasma monolayer. Phys Plasmas 28(7):073–705. https://doi.org/10.1063/5.0047649
142. Cui C, Goree J (1994) Fluctuations of the charge on a dust grain in a plasma. IEEE Trans Plasma Sci 22:151–158. https://doi.org/10.1109/27.279018
143. Praburam G, Goree J (1996) Experimental observation of very low-frequency macroscopic modes in a dusty plasma. Phys Plasmas 3(4):1212–1219. https://doi.org/10.1063/1.871745
144. Asgari H, Muniandy SV, Wong CS (2011) Stochastic dynamics of charge fluctuations in dusty plasma: a non-Markovian approach. Phys Plasmas 18(083):709. https://doi.org/10.1063/1.3626552
145. Thurston GB (1972) Viscoelasticity of human blood. Biophys J 12:1205–1217. https://doi.org/10.1016/S0006-3495(72)86156-3
146. Thurston GB (1996) Viscoelastic properties of blood and blood analogs. Adv Hemodynamics Hemorheol 1:1–30
147. Beris AN, Horner JS, Jariwala S, Armstrong MJ, Wagner NJ (2021) Recent advances in blood rheology: a review. Soft Matter 17:10 591–10 613. https://doi.org/10.1039/D1SM01212F

148. Dintenfass L (1964) Rheologic approach to thrombosis and atherosclerosis. Angiology 15(8):333–343. https://doi.org/10.1177/000331976401500802
149. Hell KM, Balzereit A, Diebold U, Bruhn HD (1989) Importance of blood viscoelasticity in arteriosclerosis. Angiology 40:539–546. https://doi.org/10.1177/000331978904000605
150. Bahloul MA, Aboelkassem Y, Laleg-Kirati T-M (2022) Human hypertension blood flow model using fractional calculus. Front Physiol 13(838):593. https://doi.org/10.3389/fphys.2022.838593
151. Mandelbrot BB, Wheeler JA (1983) The fractal geometry of nature. Am J Phys 51(3):286–287. https://doi.org/10.1119/1.13295
152. Mainster MA (1990) The fractal properties of retinal vessels: embryological and clinical implications. EYE 4(Pt 1):235–241. https://doi.org/10.1038/eye.1990.33
153. Chen Y, Zhang X, Ren L, Geng Y, Bai G (2021) Analysis of blood flow characteristics in fractal vascular network based on the time fractional order. Phys Fluids 33(041):902. https://doi.org/10.1063/5.0046622
154. Ionescu C, Lopes A, Copot D, Machado J, Bates J (2017) The role of fractional calculus in modeling biological phenomena: a review. Commun Nonlinear Sci Numer Simul 51:141–159. https://doi.org/10.1016/j.cnsns.2017.04.001
155. Avgerinos NA, Neofytou P (2019) Mathematical modelling and simulation of atherosclerosis formation and progress: a review. Ann Biomed Eng 47:1764–1785. https://doi.org/10.1007/s10439-019-02268-3
156. Cilla M, Peña E, Martínez MA (2013) Mathematical modelling of atheroma plaque formation and development in coronary arteries. J R Soc Interface 11(20130):866. https://doi.org/10.1098/rsif.2013.0866
157. Jamil DF, Saleem S, Roslan R, Al-Mubaddel FS, Rahimi-Gorji M, Issakhov A, Din SU (2021) Analysis of non-newtonian magnetic casson blood flow in an inclined stenosed artery using caputo-fabrizio fractional derivatives. Comput Methods Programs Biomed 203(106):044. https://doi.org/10.1016/j.cmpb.2021.106044
158. He S, Fataf N, Banerjee S, Sun K (2019) Complexity in the muscular blood vessel model with variable fractional derivative and external disturbances. Phys A: Stat Mech Appl 526(120):904. https://doi.org/10.1016/j.physa.2019.04.140
159. Yadeta HB, Shaw S (2023) Analysis of unsteady non-Newtonian Jeffrey blood flow and transport of magnetic nanoparticles through an inclined porous artery with stenosis using the time fractional derivative. J Appl Phys 134(10):104–701. https://doi.org/10.1063/5.0165216
160. Yao S-W, Faridi WA, Asjad MI, Jhangeer A, Inc M (2021) A mathematical modelling of a atherosclerosis intimation with atangana-baleanu fractional derivative in terms of memory function. Results Phys 27(104):425. https://doi.org/10.1016/j.rinp.2021.104425
161. Zhou Y, Song Y, Liu Z, Li W, Guo Y, Matkovic LA, Yang X, Ma R, Wan M et al (2022) The viscoelastic characteristics of in-vitro carotid plaque by kelvin-voigt fractional derivative modeling. J Biomech 141(111):210. https://doi.org/10.1016/j.jbiomech.2022.111210
162. Pérez Zerpa J, Canelas A, Sensale B, Bia Santana D, Armentano R (2015) Modeling the arterial wall mechanics using a novel high-order viscoelastic fractional element. Appl Math Model 39(16):4767–4780. https://doi.org/10.1016/j.apm.2015.04.018
163. Ramaswamy S (2010) The mechanics and statistics of active matter. Annu Rev Condens 1(1):323–345. https://doi.org/10.1146/annurev-conmatphys-070909-104101
164. Marchetti MC, Joanny JF, Ramaswamy S, Liverpool TB, Prost J, Rao M, Simha RA (2013) Hydrodynamics of soft active matter. Rev Mod Phys 85:1143–1189. https://doi.org/10.1103/RevModPhys.85.1143
165. Jülicher F, Grill SW, Salbreux G (2018) Hydrodynamic theory of active matter. Rep Prog Phys 81(076):601. https://doi.org/10.1088/1361-6633/aab6bb
166. McKinley SA, Yao L, Forest MG (2009) Transient anomalous diffusion of tracer particles in soft matter. J Rheol 53:1487–1506. https://doi.org/10.1122/1.3238546
167. Joo S, Durang X, Lee O-C, Jeon J-H (2020) Anomalous diffusion of active brownian particles cross-linked to a networked polymer: Langevin dynamics simulation and theory. Soft Matter 16:9188–9201. https://doi.org/10.1039/D0SM01200A

168. Ribeiro H, Potiguar F (2016) Active matter in lateral parabolic confinement: from subdiffusion to superdiffusion. Phys A Stat 462:1294–1300. https://doi.org/10.1016/j.physa.2016.06.139
169. Dix JA, Verkman A (2008) Crowding effects on diffusion in solutions and cells. Annu Rev Biophys 37:247–263. https://doi.org/10.1146/annurev.biophys.37.032807.125824
170. Bechinger C, Di Leonardo R, Löwen H, Reichhardt C, Volpe G, Volpe G (2016) Active particles in complex and crowded environments. Rev Mod Phys 88(045):006. https://doi.org/10.1103/RevModPhys.88.045006
171. Ellis R (2001) Macromolecular crowding: obvious but underappreciated. Trends Biochem Sci 26(10):597–604. https://doi.org/10.1016/S0968-0004(01)01938-7
172. Abbaspour L, Klumpp S (2021) Enhanced diffusion of a tracer particle in a lattice model of a crowded active system. Phys Rev E 103(052):601. https://doi.org/10.1103/PhysRevE.103.052601
173. Stone H, Stroock A, Ajdari A (2004) Engineering flows in small devices: microfluidics toward a lab-on-a-chip. Annu Rev Fluid Mech 36(1):381–411. https://doi.org/10.1146/annurev.fluid.36.050802.122124
174. Convery N, Gadegaard N (2019) 30 years of microfluidics. Micro Nano Eng 2:76–91. https://doi.org/10.1016/j.mne.2019.01.003
175. Li X-X, Yin Z, Jian Y-J, Chang L, Su J, Liu Q-S (2012) Transient electro-osmotic flow of generalized maxwell fluids through a microchannel. J Non-Newton Fluid Mech 187–188:43–47. https://doi.org/10.1016/j.jnnfm.2012.09.005
176. Wang S, Zhao M, Li X (2014) Transient electro-osmotic flow of generalized maxwell fluids in a straight pipe of circular cross section. Open Phys 12:445–451. https://doi.org/10.2478/s11534-014-0463-x
177. Sasmal C (2022) Fluid viscoelasticity suppresses chaotic convection and mixing due to electrokinetic instability. Phys Fluids 34(082):011. https://doi.org/10.1063/5.0099481
178. Feng C, Si X, Cao L, Zhu B (2021) The slip flow of generalized maxwell fluids with time-distributed characteristics in a rotating microchannel. Appl Math Lett 120(107):260. https://doi.org/10.1016/j.aml.2021.107260
179. Tian K, An S, Zhao G, Ding Z (2022) Two-dimensional electromagnetohydrodynamic (EMHD) flows of fractional viscoelastic fluids with electrokinetic effects. Nanomaterials (Basel) 12:3335. https://doi.org/10.3390/nano12193335
180. Cao L, Zhang P, Li B, Zhu J, Si X (2021) Numerical study of rotating electro-osmotic flow of double layers with a layer of fractional second-order fluid in a microchannel. Appl Math Lett 111(106):633. https://doi.org/10.1016/j.aml.2020.106633
181. Yang X, Qi H, Jiang X (2018) Numerical analysis for electroosmotic flow of fractional maxwell fluids. Appl Math Lett 78:1–8. https://doi.org/10.1016/j.aml.2017.10.012
182. Liu Y, Zhang H & Jiang X 2021 Fast evaluation for magnetohydrodynamic flow and heat transfer of fractional oldroyd-b fluids between parallel plates. Z Angew Math Mech 101(10):e202100 042. https://doi.org/10.1002/zamm.202100042

Appendix
Solutions to Problems

A.1 Solutions to Chap. 2

1. Application of the Laplace transform with respect to t gives

$$s^{\alpha}\bar{u}(z,s) = \nu\frac{d^2\bar{u}}{dz^2}, \quad 0 < z < \infty, \tag{A.1}$$

$$\bar{u}(0,s) = U\bar{f}(s), \quad \bar{u}(z,s) \to 0 \quad \text{as} \quad z \to \infty. \tag{A.2}$$

Using the Fourier sine transform with respect to z yields

$$\bar{U}_s(k,s) = \left(\sqrt{\frac{2}{\pi}}\nu U\right)\frac{k\bar{f}(s)}{s^{\alpha} + \nu k^2}. \tag{A.3}$$

The inverse Fourier sine transform of (A.3) leads to the solution

$$\bar{u}(z,s) = \left(\frac{2}{\pi}\nu U\right)\bar{f}(s)\int_0^{\infty}\frac{k\sin kz}{s^{\alpha} + \nu k^2}dk, \tag{A.4}$$

and the inverse Laplace transform gives the solution for the velocity field is,

$$u(z,t) = \left(\frac{2}{pi}\nu U\right)\int_0^{\infty} k\sin kz dk \times \int_0^t f(t-\tau)\tau^{\alpha-1}E_{\alpha,\alpha}(-\nu k^2\tau^{\alpha})d\tau. \tag{A.5}$$

When $f = e^{i\omega t}$, the solution of the fractional Stokes problem is found,

$$u(z,t) = \left(\frac{2}{pi}\nu U\right)e^{i\omega t}\int_0^{\infty} k\sin kz dk \times \int_0^t e^{-i\omega\tau}\tau^{\alpha-1}E_{\alpha,\alpha}(-\nu k^2\tau^{\alpha})d\tau, \tag{A.6}$$

H. Wilson et al., *Viscoelastic Subdiffusive Flows*, Fluid Mechanics and Its Applications 138, https://doi.org/10.1007/978-981-96-0135-6

and for $f(t) = 1$, the solution of the fractional Rayleigh problem follows,

$$u(z,t) = \left(\frac{2}{pi}\nu U\right)\int_0^{\infty} k\sin kz dk \times \int_0^t \tau^{\alpha-1}E_{\alpha,\alpha}(-\nu k^2\tau^{\alpha})d\tau. \tag{A.7}$$

2. Use formula (2.22) and (2.25).

A.2 Solutions to Chap. 3

1. a. The governing equations of viscoelastic fluid include the continuity equation, Cauchy momentum equation, and stress balance equation. The upper-convected Maxwell (UCM) model is chosen for the stress. The governing equations are as follows:

$$\begin{aligned} &\nabla\cdot\boldsymbol{u} = 0, \qquad \rho\left[\frac{\partial\boldsymbol{u}}{\partial t} + \boldsymbol{u}\cdot\nabla\boldsymbol{u}\right] = -\nabla p + \eta\nabla\tau, \\ &\lambda\left[\frac{\partial\tau}{\partial t} + (\boldsymbol{u}\cdot\nabla)\tau - (\nabla\boldsymbol{u})^T\tau - \tau(\nabla\boldsymbol{u})\right] = \nabla\boldsymbol{u} + (\nabla\boldsymbol{u})^T - \tau, \end{aligned} \tag{A.8}$$

where $\boldsymbol{u} = (u, v)$ is the velocity vector, and ρ, p η, λ, and $\tau = [T_{ij}]$ are the density, pressure, viscosity, (microstructural) relaxation time, and polymeric (symmetric) stress tensor, respectively. By choosing length scale as half of channel height h, velocity scale as the maximum channel velocity U (or average wall speed), time scale as the inverse of overall shear-rate h/U. Substituting $(\bar{x}, \bar{y}) = (x, y)/h$, $(\bar{u}, \bar{v}) = (u, v)/U$, $\bar{t} = t/(h/U)$ into governing equations (A.8) and omitting the bar over the quantities, the following dimensionless governing equations are obtained,

$$\begin{aligned} &\nabla\cdot\boldsymbol{u} = 0, \qquad Re\left[\frac{\partial\boldsymbol{u}}{\partial t} + \boldsymbol{u}\cdot\nabla\boldsymbol{u}\right] = -\nabla p + \nabla\tau, \\ &We\left[\frac{\partial\tau}{\partial t} + (\boldsymbol{u}\cdot\nabla)\tau - (\nabla\boldsymbol{u})^T\tau - \tau(\nabla\boldsymbol{u})\right] = \nabla\boldsymbol{u} + (\nabla\boldsymbol{u})^T - \tau, \end{aligned} \tag{A.9}$$

where $Re = \rho hU/\eta$ is the Reynolds number—measures the ratio of inertial to viscous forces—and $We = \lambda U/h$ is the Weissenberg number—measures the ratio of elastic to viscous forces.

b. Under steady ($\partial(.)/\partial t = 0$), unidirectional (velocity is only along the x-direction), and fully developed ($\partial(.)/\partial x = 0$) assumptions, the flow variables are functions of y. Therefore, the velocity has the form $(u, v) = (u^0(y), 0)$, and other variables vary along the y-direction alone. Here the superscript 0 over a quantity denotes the steady state. We get a simplified set of equations by applying these assumptions to (A.9). In particular, the continuity equation trivially satisfies, and the momentum and stress balance equations read

$$\frac{\partial T^0_{12}}{\partial y} = 0, \quad \frac{\partial T^0_{22}}{\partial y} = 0, \quad T^0_{11} = 0, \quad T^0_{12} = \frac{\partial u^0}{\partial y}, \quad T^0_{22} = 2We\left(\frac{\partial u^0}{\partial y}\right)^2. \tag{A.10}$$

The first and fourth equations give $\partial^2 u^0/\partial y^2 = 0$. Since $u^0(\pm 1) = \pm 1$, we get a linear velocity profile $u^0(y) = y$, which gives $p^0 = 0,\ T^0_{11} = 0,\ T^0_{12} = 1,\ T^0_{22} = 2We$.

2. The linear stability analysis is executed by applying the classical normal-mode analysis. We substitute the flow variables $X = (u, v, p, T_{11}, T_{12}, T_{22})$ in the governing equations (A.9) as a sum of base flow $X^0 = (u^0, v^0 = 0, p^0 = 0, T^0_{11}, T^0_{12}, T^0_{22})$ and the perturbations $\tilde{X} = (\tilde{u}, \tilde{v}, \tilde{p}, \tilde{T}_{11}, \tilde{T}_{12}, \tilde{T}_{22})$, i.e. $X = X^0 + \tilde{X}$. Subtracting the base flow equations (A.10) from the resulting equations and neglecting the nonlinear terms of the perturbations, we get the linearized perturbation equations.
Next, we assume the normal mode solution for the perturbation variables- $\tilde{X} = \epsilon\hat{X}(y)e^{ik(x-ct)}$, where k, c, and ϵ are the stream-wise wave number, complex phase velocity, and a small number, respectively, and $\hat{X}(y) = (\hat{u}, \hat{v}, \hat{p}, \hat{T}_{11}, \hat{T}_{12}, \hat{T}_{22})$ is the amplitude of the perturbation. Substituting the normal mode solution into the linearized perturbation equations and collecting the $\epsilon e^{ik(x-ct)}$ terms, we obtain the following set of equations,

$$ik\hat{u} + \frac{d\hat{v}}{dy} = 0,$$

$$Re\Big[(-i\,k\,c)\hat{u} + iku^0\hat{u} + u^0_y\hat{v}\Big] = -ik\hat{p} + ik\hat{T}_{11} + \frac{\hat{T}_{12}}{dy},$$

$$Re\Big[(-i\,k\,c)\hat{v} + iku^0\hat{v}\Big] = -\frac{d\hat{p}}{dy} + ik\hat{T}_{12} + \frac{d\hat{T}_{22}}{dy},$$

$$We\Big[(-i\,k\,c)\hat{T}_{11} + iku^0\hat{T}_{11} - T^0_{11y}\hat{v} - 2ikT^0_{11}\hat{u} - 2ikT^0_{12}\hat{v}\Big] = 2ik\hat{u} - \hat{T}_{11},$$

$$We\Big[(-i\,k\,c)\hat{T}_{12} + iku^0\hat{T}_{12} + T^0_{12y}\hat{v} - u^0_y\hat{T}_{11} - T^0_{11}\frac{d\hat{u}}{dy} - ikT^0_{22}\hat{v}\Big] = \frac{d\hat{u}}{dy} + ik\hat{v} + \hat{T}_{12},$$

$$We\Big[(-i\,k\,c)\hat{T}_{22} + iku^0\hat{T}_{22} + T^0_{22y}\hat{v} - 2T^0_{12}\frac{d\hat{u}}{dy} - 2u^0_y\hat{T}_{12} - 2T^0_{22}\frac{d\hat{v}}{dy}\Big] = 2\frac{d\hat{v}}{dy} - \hat{T}_{22},$$

where $(u^0, T^0_{11}, T^0_{12}, T^0_{22}) = (y, 0, 1, 2We)$ and $(u^0_y, T^0_{11y}, T^0_{12y}, T^0_{22y}) = (1, 0, 0, 0)$. Note that the pressure $\hat{p}$, stress components $\hat{T}_{11}$, $\hat{T}_{12}$, $\hat{T}_{22}$, and their derivatives can be explicitly expressed in terms of the amplitude of perturbed stream-function $\hat{\psi}$, defined as $\hat{u} = \frac{d\hat{\psi}}{dy}$ and $\hat{v} = -ik\hat{\psi}$. Substituting the variables $\hat{p}, \hat{T}_{11}, \hat{T}_{12}, \hat{T}_{22}$ in terms of $\hat{\psi}$ and its derivatives in y-momentum equation and simplifying, we get the Orr–Sommerfeld equation for viscoelastic plane Couette flow with UCM model:

$$a_4 \frac{d^4\hat{\psi}}{dy^4} + a_3 \frac{d^3\hat{\psi}}{dy^3} + a_2 \frac{d^2\hat{\psi}}{dy^2} + a_1 \frac{d\hat{\psi}}{dy} + a_0\hat{\psi} = 0,$$

where the coefficients a_j are

$$\begin{aligned}
a_0 =& Rek^2(y-c) + \frac{24ik^3We^4}{\mathcal{R}^4} + \frac{4ik^3We^2}{\mathcal{R}^2} + \frac{k^3(1+2We^2)}{\mathcal{R}} \\
&+ \frac{6k^3We^2(1+2We^2)}{\mathcal{R}^3} \\
a_1 =& -\frac{24k^2We^3}{\mathcal{R}^4} + \frac{4k^2We^3}{\mathcal{R}^2} + \frac{2ik^2We}{\mathcal{R}} + \frac{12ik^2We^3}{\mathcal{R}^3} - \frac{2k^2We}{\mathcal{R}^2} \\
a_2 =& Re(c-y) - \frac{6kWe^2}{\mathcal{R}^3} - \frac{2k}{\mathcal{R}} - \frac{2kWe^2}{\mathcal{R}} \\
a_3 =& \frac{2We(1-iR)}{\mathcal{R}^2}, \\
a_4 =& \frac{1}{k\mathcal{R}}, \\
\mathcal{R} =& i + ckWe - kWey.
\end{aligned}$$

3. a. All details are the same as Problem 1(a).
   b. The plane Poiseuille flow (PPF) is driven by the constant longitudinal pressure gradient, i.e. $\partial p^0/\partial x =$ constant. The steps are the same as explained in Problem 1(b). The continuity equation is trivially satisfied under the steady, unidirectional, and fully developed base flow assumptions. The momentum and stress balance equations read

$$\frac{\partial T^0_{12}}{\partial y} = \frac{\partial p^0}{\partial x}, \quad \frac{\partial T^0_{22}}{\partial y} = \frac{\partial p^0}{\partial y}, \quad T^0_{11}=0, \quad T^0_{12} = \frac{\partial u^0}{\partial y}, \quad T^0_{22} = 2We\left(\frac{\partial u^0}{\partial y}\right)^2. \tag{A.11}$$

   The first and fourth equations give $\partial^2 u^0/\partial y^2 = \partial p^0/\partial x =$ constant. Since plates are rigid, no-slip boundary conditions on velocity are applied, i.e. $u^0(y=\pm 1) = 0$. Solving the equation with no-slip conditions, a parabolic velocity profile $u^0(y) = 1 - y^2$ is obtained. Similarly solving remaining equations with $u^0 = 1 - y^2$, we get the pressure as $p^0 = -2x + 8Wey^2$ and stress components as $T^0_{11} = 0,\ T^0_{12} = -2y,\ T^0_{22} = 8We\, y^2$.
4. Similar to the Problem 2, the Orr–Sommerfeld equation for the viscoelastic plane Poiseuille flow is given as

$$a_4 \frac{d^4\hat{\psi}}{dy^4} + a_3 \frac{d^3\hat{\psi}}{dy^3} + a_2 \frac{d^2\hat{\psi}}{dy^2} + a_1 \frac{d\hat{\psi}}{dy} + a_0\hat{\psi} = 0,$$

where

$$
\begin{aligned}
a_0 =& Re(-2 + k^2(1 - y^2 - c)) + \frac{k^3}{\mathcal{R}} - \frac{6k^2 We}{\mathcal{R}^2} - \frac{6ik^2 We}{\mathcal{R}} + \frac{28ik We^2}{\mathcal{R}^2},\\
&+ \frac{32k We^2}{\mathcal{R}} + \frac{24k^3 We^2 y^2}{\mathcal{R}^3} + \frac{16ik^3 We^2 y^2}{\mathcal{R}^2} + \frac{8k^3 We^2 y^2}{\mathcal{R}},\\
&- \frac{304ik^2 We^3 y^2}{\mathcal{R}^3} - \frac{208k^2 We^3 y^2}{\mathcal{R}^2} + \frac{384ik^3 We^4 y^4}{\mathcal{R}^4} + \frac{192k^3 We^4 y^4}{\mathcal{R}^3},\\
a_1 =& \frac{4k^2 Wey}{\mathcal{R}^2} - \frac{4ik^2 Wey}{\mathcal{R}} - \frac{80k We^2 y}{\mathcal{R}^3} + \frac{48ik We^2 y}{\mathcal{R}^2},\\
&+ \frac{16k We^2 y}{\mathcal{R}} + \frac{192k^2 We^3 y^3}{\mathcal{R}^4} - \frac{96ik^2 We^3 y^3}{\mathcal{R}^3} - \frac{32k^2 We^3 y^3}{\mathcal{R}^2},\\
a_2 =& Re(-1 + c + y^2) - \frac{2k}{\mathcal{R}} + \frac{2We}{\mathcal{R}^2} + \frac{2iWe}{\mathcal{R}} - \frac{24k We^2 y^2}{\mathcal{R}^3} - \frac{8k We^2 y^2}{\mathcal{R}},\\
a_3 =& \frac{4Wey(-1 + i\mathcal{R})}{\mathcal{R}^2},\\
a_4 =& 1/(k\mathcal{R}),\\
\mathcal{R} =& i - kWe(1 - y^2 - c).
\end{aligned}
$$

A.3 Solutions to Chap. 4

1. The standard Riemannian metric on $\mathbb{R}^d$ is an example of a complete metric. For the non-complete example, equip $\mathbb{R}^d$ with any explicit, sufficiently rapidly decaying Riemannian metric, for example, $g_{ij}(x)$, with

$$
g_{ij}(x) = \delta_{ij} e^{-2\|x\|^2}, \quad \text{with} \quad \|x\|^2 = (x^1)^2 + \ldots + (x^d)^2. \tag{A.12}
$$

Note that each matrix entry $g_{ij}(x)$ is a C^∞ function of $x \in \mathbb{R}^d$ and also the matrix $g_{ij}(x)$ is positive definite for every $x \in \mathbb{R}^d$ (since it is a positive multiple of the identity matrix); hence the formula indeed gives a Riemannian metric on $\mathbb{R}^d$. To prove that this metric is not complete, consider the sequence of points $p_1, p_2, \ldots$ with $p_j = (j, 0, \ldots, 0)$. For any integers $1 \le j \le k$, we have.

$$
d(p_j, p_k) \le e^{-j}. \tag{A.13}
$$

Indeed, consider the C^∞ curve $c : [j, k] \to M, \ c(t) = (t, 0, \ldots, 0)$. This is a curve from p_j to p_k and its length with respect to the Riemannian metric (A.12) is,

$$L(c) = \int_j^k \sqrt{\langle \dot{c}(t), \dot{c}(t) \rangle} dt = \int_j^k \sqrt{e^{-2t^2}} dt = \int_j^k e^{-t^2} dt < \int_j^\infty e^{-t} dt = e^{-j}. \tag{A.14}$$

(Here we used the fact that $t^2 > t$ for all $t > j \geq 1$). This proves that (A.14) holds, and (A.14) in turn implies that $p_1, p_2, \ldots$ is a Cauchy sequence in $\mathbb{R}^d$ equipped with the Riemannian metric (A.12). On the other hand the sequence $p_1, p_2, \ldots$ does not converge to any point in $\mathbb{R}^d$. Hence $\mathbb{R}^d$ equipped with the Riemannian metric (A.12) is not complete.

2. Let us first note that this is not an immediate consequence of the following fact from basic point set topology: Every closed subset of a complete metric space is itself a complete metric space. Namely, in that statement it is understood that the subset is endowed with the metric, which is simply the restriction of the metric on the larger space. This is not the case in our situation; we typically have $d_M(p, q) \neq d_N(p, q)$ for $p, q \in M$. However, as weÕll see, the completeness of M is still easy to prove...
   Let $p_1, p_2, \ldots$ be a Cauchy sequence in (M, d_M). Then, we have that $d_N(p_j, p_k) \leq d_M(p_j, p_k)$ for any j, k; hence $p_1, p_2, \ldots$ is also a Cauchy sequence in (N, d_N). Therefore, since N is complete there exists a (unique) limit point $p = \lim_{j\to\infty} p_j$ in N. Recall that this limit relation by definition means that,
$$\lim_{j\to\infty} d_N(p_j, p) = 0. \tag{A.15}$$
   Since M is closed in N we have $p \in M$. But since M is an embedded submanifold of N, the topology of M equals the subspace topology of M as a subset of N; therefore $\lim_{j\to\infty} d_N(p_j, p) = 0$ implies $\lim_{j\to\infty} d_M(p_j, p) = 0$ or $p = \lim_{j\to\infty} p_j$ in M. This proves that M is complete.
3. a. This continuity is clear from
$$|d(p, q) - d(p', q')| \leq d(p, p') + d(q, q'), \quad \text{for all} p, p', q, q' \in X. \tag{A.16}$$
   b. Take any $q \in X$. If $d(p, q) < r$ then setting $s = r - d(p, q) > 0$ we have $B_s(q) \subset B_r(p)$ (by triangle inequality) and so $q \notin \partial B_r(p)$. If $d(p, q) > r$ then setting $s = d(p, q) - r > 0$ we have $B_s(q) \cap B_r(p) = \emptyset$ (again by the triangle inequality) and so $q \notin \partial B_r(p)$. This proves the stated inclusion. The set $\partial B_r(p)$ is closed since the boundary of any set (in any topological space) is closed. The fact that the set $\{q \in X | d(p, q) = r\}$ is closed is an immediate consequence of the fact that the metric d is continuous.
   Next, assume that (X, d) is a Riemannian manifold. Take any $q \in X$ with $d(p, q) = r$. By the definition of "d", there exists a sequence of C^∞ curves $\gamma_1, \gamma_2, \ldots$ on X such that each γ_j starts at p and ends at q and $\ell_j = L(\gamma_j) < r + j^{-1}$; also $\ell_j \geq r$. We may assume that each γ_j is parametrized by arc length, and has domain $[0, \ell_j]$ where $\ell_j = L(\gamma_j)$. Take $J \in \mathbb{R}^+$ so large that $J^{-1} < r$, and for each $j \geq J$ set,

$$q_j = \gamma_j(r - j^{-1}).$$

Note that $\gamma_j|_{[0,r-j^{-1}]}$ is a curve of length $r - j^{-1}$ from p to q_j; hence $d(p, q_j) \le r - j^{-1}$ and $q_j \in B_r(p)$. On the other hand $\gamma_j|_{[r-j^{-1},\ell_j]}$ is a curve of length

$$\le \ell_j - (r - j^{-1}) < (r + j^{-1}) - (r - j^{-1}) = 2j^{-1},$$

from q_j to q; hence $d(q_j, q) < 2j^{-1}$. Hence $q_j \to q$ as $j \to \infty$. This shows that q is in the closure of the set $B_r(p)$. Also $q \notin B_r(p)$ since $d(p, q) = r$. Hence $q \in \partial B_r(p)$. We have thus proved that $\{q \in X : d(p, q) = r\} \subset B_r(p)$.

c. We have

$$d(p, q) \le d(p, p_0) + d(p_0, q)$$

by the triangle inequality; hence it now suffices to prove the opposite inequality. Let $\gamma : [a, b] \to X$ be any C^∞ curve with $\gamma(a) = p$ and $\gamma(b) = q$. Since $t \to d(p, \gamma(t))$ is a continuous function of t, and $d(p, \gamma(a)) = 0, d(p, \gamma(b)) = d(p, q) > r$, there must exist some $t_0 \in (a, b)$ such that $d(p, \gamma(t_0)) = r$. Using part (b), we then have $\gamma(t_0) \in \partial B_r(p)$, and hence because of the way p_0 was chosen,

$$d(\gamma(t_0), q) \ge d(p_0, q).$$

Therefore,

$$L(\gamma) = L(\gamma_{[a,t_0]}) + L(\gamma_{[t_0,b]}) \ge d(p, \gamma(t_0)) + d(\gamma(t_0), q) \ge r + d(p_0, q).$$

Since this is true for every C^∞ curve from p to q, we have,

$$d(p, q) \ge r + d(p_0, q) = d(p, p_0) + d(p_0, q),$$

and the proof is complete.

A.4 Solutions to Chap. 5

1. We utilize the finite difference θ-method (developed in Sect. 5.2.1) to analyze the stability of the model. To achieve this, we will use a rectangular domain of size $w^h \times w^\tau$, divided into equal intervals $h = \frac{w^h}{M}$ and $\tau = \frac{w^\tau}{N}$, for the x-axis and the t-axis, respectively. And $x_n = nh, n = 1, 2, \ldots, M$; $t_k = k\tau, k = 1, 2, \ldots, N$ and rewrite the model at points (t_k, x_n) as,

$$z_{ttt}(t_k, x_n) + z_{tt}(t_k, x_n) + {}_0^C D_t^{\alpha}(t_k, x_n) + z(t_k, x_n) = z_{xx}(t_k, x_n) + f(t_k, x_n). \tag{A.17}$$

The first-order difference technique for the Caputo derivative, ${}_0^C D_t^{\alpha}(t_k, x_n)$ is defined as,

$${}_0^C D_t^{\alpha}(t_k, x_n) \approx s_{\alpha,\tau} \sum_{j=0}^{k} \psi_j^{(\alpha)} (z_n^{k-j+1} - z_n^{k-j}), \tag{A.18}$$

where $\psi_j^{(\alpha)} = (j+1)^{1-\alpha} - j^{1-\alpha}$, $s_{\alpha,\tau} = \frac{\tau^{-\alpha}}{\Gamma(2-\alpha)}$. To discretize the remaining terms, we utilize the Taylor series expansion,

$$z_{xx}(t_k, x_n) \approx \frac{1}{h^2}\{\theta[z_{n+1}^{k+1} - 2z_n^{k+1} + z_{n-1}^{k+1}] + (1-\theta)[z_{n+1}^{k} - 2z_n^{k} + z_{n-1}^{k}]\}, \tag{A.19}$$

$$z_{tt}(t_k, x_n) \approx \frac{z_n^{k+1} - 2z_n^k + z_n^{k-1}}{\tau^2}, \tag{A.20}$$

$$z_{ttt}(t_k, x_n) \approx \frac{z_n^{k+2} - 3z_n^{k+1} + 3z_n^k - z_n^{k-1}}{\tau^3}. \tag{A.21}$$

Equations (A.19)–(A.21) can be utilized to create a method of difference scheme as follows,

$$\begin{aligned}
&\frac{z_n^{k+2} - 3z_n^{k+1} + 3z_n^k - z_n^{k-1}}{\tau^3} + \frac{z_n^{k+1} - 2z_n^k + z_n^{k-1}}{\tau^2} + s_{\alpha,\tau}\sum_{j=0}^{k} \psi_j^{(\alpha)} (z_n^{k-j+1} - z_n^{k-j}), \\
&- \frac{1}{h^2}\{\theta[z_{n+1}^{k+1} - 2z_n^{k+1} + z_{n-1}^{k+1}] + (1-\theta)[z_{n+1}^{k} - 2z_n^{k} + z_{n-1}^{k}]\}, \\
&+ \theta z_n^{k+1} + (1-\theta) z_n^k = \theta f_n^{k+1} + (1-\theta) f_n^k,
\end{aligned} \tag{A.22}$$

and $z_n^0 = \sigma_1(x_n)$, $\frac{z_n^1 - z_n^0}{\tau} = \sigma_2(x_n)$, $\frac{z_n^2 - 2z_n^1 + z_n^0}{\tau^2} = \sigma_3(x_n)$, $z_0^k = z_M^k = 0, 0 \le k \le N$. To perform the Von-Neumann analysis for formula (A.22), assume,

$$z_n^k = \lambda^k e^{in\phi} \tag{A.23}$$

Substituting equation (A.23) in formula (A.22) and using the initial conditions $(f_0^1, f_0^2 = 0)$, we get,

$$
\frac{\lambda^{k+2}e^{in\phi}-3\lambda^{k+1}e^{in\phi}-\lambda^{k-1}e^{in\phi}}{\tau^3}+\frac{\lambda^{k+1}e^{in\phi}-2\lambda^k e^{in\phi}+\lambda^{k-1}e^{in\phi}}{\tau^2}
$$
$$
+s_{\alpha,\tau}\sum_{j=0}^{k}\psi_j^{(\alpha)}(\lambda^{k-j+1}e^{in\phi}-\lambda^{k-j}e^{in\phi})
$$
$$
-\frac{1}{h^2}\{\theta\lambda^{k+1}(e^{i(n+1)\phi}-2e^{in\phi}+e^{i(n-1)\phi})
$$
$$
+\theta\lambda^{k+1}e^{in\phi}+(1-\theta)\lambda^k e^{in\phi}
$$
$$
+(1-\theta)\lambda^k(e^{i(n+1)\phi}-2e^{in\phi}+e^{i(n-1)\phi})=f_n^k+\theta(f_n^{k+1}-f_n^k). \quad (A.24)
$$

Taking $k=1$ and $n=0$, we then obtain,

$$
\frac{\lambda^3-3\lambda^2+3\lambda-1}{\tau^3}+\frac{\lambda^2-2\lambda+1}{\tau^2}+s_{\tau,\alpha}[(\lambda-1)(2^{1-\alpha}-1)
$$
$$
+(\lambda^2-\lambda)]-\frac{\theta\lambda^2}{h^2}(e^{i\varphi}-2+e^{-i\phi})+\theta\lambda^2
$$
$$
+(1-\theta)\lambda-\frac{(1-\theta)\lambda}{h^2}(e^{i\phi}-2+e^{-i\phi})=0. \quad (A.25)
$$

Rewriting formula (A.25) as $P(\lambda)=a\lambda^3+b\lambda^2+c\lambda+d$, where

$$
\begin{aligned}
a &= 1,\\
b &= \tau^3\left(-\frac{3}{\tau^3}+\frac{1}{\tau^2}+s_{\alpha,\tau}+4\frac{\theta}{h^2}\sin^2\frac{\phi}{2}+\theta\right),\\
c &= \tau^3\left(\frac{3}{\tau^3}-\frac{2}{\tau^2}+2(2^{-\alpha}-1)s_{\alpha,\tau}+4\frac{1-\theta}{h^2}\sin^2\frac{\phi}{2}+1-\theta\right),\\
d &= \tau^3\left(-\frac{1}{\tau^3}+\frac{1}{\tau^2}-(2^{1-\alpha}-1)s_{\alpha,\tau}\right).
\end{aligned} \quad (A.26)
$$

Taking,

$$
\begin{aligned}
p &= \sqrt{27d^2+4b^3d-18bdc+4c^3-(bc)^2},\\
s &= \frac{p}{6\sqrt{3}}-\frac{27d-9bc+2b^3}{54},
\end{aligned} \quad (A.27)
$$

we have the following roots of $P(\lambda)=0$,

$$
\begin{aligned}
\lambda_1 &= s^{1/3}+\frac{b^2-3c}{9\sqrt{s}}-\frac{b}{3},\\
\lambda_2 &= \lambda_1\left(-\frac{1}{2}+i\frac{\sqrt{3}}{2}\right),
\end{aligned}
$$

$$\lambda_3 = \lambda_1 \left(-\frac{1}{2} - i\frac{\sqrt{3}}{2} \right). \tag{A.28}$$

From Eq. (A.28), check that $|\lambda_1| = |\lambda_2| = |\lambda_3| = r$. We can run a MATLAB code to find that the maximum value of "r" is at $\tau = h = 0.5, \phi = \pi, \alpha = 0, \theta = 1$ and this maximum is $r = 0.2132$. Since the magnitude of the roots are less than 1, the finite difference scheme (A.22) is unconditionally stable.

2. The time fractional Burgers hierarchy is determined from the following equation,

$$D_t^{\alpha} u + \mu D_x (D_x + u)^s u = 0, \quad s = 0, 1, 2, \dots, \tag{A.29}$$

where μ is an arbitrary constant and $0 < \alpha < 1$ is the fractional order of derivative with respect to time. The time fractional 5th order equation (TFFB) is found by setting $s = 4$ in Eq. (A.29) and expanding,

$$\begin{aligned} &D_t^{\alpha} u + \mu \left(u_{5x} + 10u_{2x}^2 + 15u_x u_{3x} 5uu_{4x} + 15u_x^3 + 50uu_x u_{2x} \right. \\ &\left. +10u^2 u_{3x} + 30u^2 u_x^2 + 10u^3 u_{2x} + 5u^4 u_x \right) = 0, \quad 0 < \alpha < 1, \end{aligned} \tag{A.30}$$

where $u_{ix} = \frac{\partial^i u}{\partial x^i}$, $i = 2, \dots, 5$. Let us consider $\psi(x, t)$ as the wave function for a $(1 + 1)-$dimensional system and sinusoidal wave form can be represented as,

$$\psi(x, t) = \text{Re}\{Ae^{i(\omega t - kx)}\}, \tag{A.31}$$

where A is known as the complex amplitude and the parameters $(\omega, k) \in \mathbb{C}$ satisfy the dispersion relation. The linear form of TFFB is given by,

$$D_t^{\alpha} u_t + \mu u_{5x} = 0. \tag{A.32}$$

The dispersion relation corresponding to equation (A.32) is,

$$(i\omega)^{\alpha} + \mu(-ik)^5 = 0. \tag{A.33}$$

where $\text{Re}(\omega) = \cos((\frac{1}{\alpha} - 1)\,\frac{\pi}{2})\mu^{\frac{1}{\alpha}} k^{\frac{5}{\alpha}}$. The phase velocity is given by,

$$v_p = \text{Re}\left(\frac{\omega}{k}\right) = \cos(\left(\frac{1}{\alpha} - 1\right)\frac{\pi}{2})\mu^{\frac{1}{\alpha}} k^{\frac{5}{\alpha} - 1}. \tag{A.34}$$

Similarly, the group velocity is given by

$$v_g = \text{Re}\left(\frac{\partial\omega}{\partial k}\right) = \frac{5}{\alpha}\cos(\left(\frac{1}{\alpha} - 1\right)\frac{\pi}{2})\mu^{\frac{1}{\alpha}} k^{\frac{5}{\alpha} - 1}. \tag{A.35}$$

Index

H. Wilson et al., *Viscoelastic Subdiffusive Flows*, Fluid Mechanics and Its Applications 138, https://doi.org/10.1007/978-981-96-0135-6

The manufacturer's authorised representative in the EU is Springer Nature Customer Service Centre GmbH, Europaplatz 3, 69115 Heidelberg, Germany. If you have any concerns regarding our products, please contact ProductSafety@springernature.com

Printed and bound by CPI Group (UK) Ltd, Croydon, CR0 4YY
01/10/2026
02225934-0001